for all the curious girls

for all the curious girls

Stories of Wonder, Exploration, and Scientific Discovery

A SOWIF ANTHOLOGY

Introduction by Dr. Kelli Hendrickson, Founder

Illustrated by Jane Shore

SWP

SHE WRITES PRESS

Published in 2025 by
She Writes Press, an imprint of The Stable Book Group

32 Court Street, Suite 2109
Brooklyn, NY 11201
https://shewritespress.com
Library of Congress Control Number: 2025915277
ISBN: 978-1-64742-990-4
eISBN: 978-1-64742-991-1

Interior Designer: Tabitha Lahr
Editor-in-Chief: Kelli Hendrickson
Editors: Karen Flack, Elizabeth Gregorio, Beverley McKeon, and Christina Rice
Illustrator: Jane Shore
Additional Illustration Editors: Roni Goldshmid, Nicole Xu

Printed in the United States

Names and identifying characteristics have been changed to protect the privacy of certain individuals.

To all of the curious girls who came before us,
And all who are yet to follow.

Contents

Preface

This anthology was created with one purpose in mind: to introduce the next generation of curious girls[1] to the fascinating world of science. It is written for middle-grade readers (ages eight to twelve) who are curious about science and working to figure out their role in their world. Whether you're the parent nurturing a girl's curiosity or a fellow scientist eager to pass the torch, this collection is for you.

The idea for this anthology was sparked by a simple question I posed at a women's networking lunch during a fluid dynamics conference: "What stories can we tell girls and young women to encourage them to join us in this field?" In those that answered this question, I found a strong and resilient group of women who helped create this book and, by doing so, helped me found the Stories of Women in Fluids (SOWIF) Initiative.*

This question also led to a powerful realization—that sharing our stories can be one of the most effective ways to inspire

1. Throughout this anthology, we use the terms "girl" and "woman" to encompass all femme-identifying scientists and young adults.

and engage. The women who contributed to this anthology have not only shared their science but have also opened a window into their personal lives, blending science communication with personal narrative.

For many of these women, science is more than just a career; it's an integral part of who they are. Through this collection, our goal is to encourage young women to see themselves in the same light—as future scientists and explorers who can change the world.

This anthology focuses on fluid dynamics, a field of science that touches nearly everything that flows around and within us yet doesn't get much attention in early science classes. Fluid dynamics governs the air we breathe, the water we swim in, the design of airplanes, and even the beating of our hearts. Simply put, it's the study of how liquids and gases move, and it's something we experience every day, whether we're aware of it or not.

To parents reading this: We encourage you to place this book in the hands of your children. The world needs more curious girls in science, technology, engineering, and mathematics (STEM). Let them explore the stories of today's women in science. By sharing these stories, you are giving them modern role models they can relate to and are introducing them to the wonders of fluid dynamics. Nurture their curiosity and encourage their questions. They'll see that science is for everyone—no matter their gender, culture, or background.

To the fellow scientists reading this: We're sure you understand how vital age-accessible science communication and outreach is for middle graders. We created this genre of science storytelling to blend science communication with personal narrative, hoping to make a stronger connection with those that read about our science. We invite you to join us in this initiative. Share your stories, your science, and your passion with the next generation of curious girls who are eager to learn.

Let's make a lasting impact. Together, we can create and inspire a more diverse, inclusive, and exciting future for STEM where science thrives because it reflects the diversity of those who practice it.

With gratitude and anticipation,
Kelli Hendrickson
Founder and Principal Instigator of the Stories of Women in Fluids Initiative

---

* The Stories of Women in Fluids Initiative (storiesofwomeninfluids.org) aims to expand beyond this book. Through future anthologies, mentorship, and outreach, we are working to foster a thriving and inclusive community that celebrates and supports women in science.

Introduction

Dear Curious Girl,

What are you curious about?

Have you ever lain in a meadow, watching the clouds dance across the sky, and wondered how the wind shapes the clouds? Or sat on a bumpy airplane ride, wishing the pilot could make the ride smoother?

The invisible air around us is always moving. Its movement pushes the leaves, sways the grass, and shapes the clouds above.

As you wander through the zoo, do you wonder how fish and seals swim as effortlessly as we walk? Have you tried to swim faster without working harder and wondered how it's done? Water flows all around us; its motion and how we can move through it is full of mysteries waiting to be discovered.

The world is full of flowing wonders—of air, water, even your blood—that are waiting to be explored. And who better to uncover their secrets than someone as curious as you?

Inside this book, you'll find stories about girls who, just like you, wanted to understand the why and how of the world around them. From watching waves crash on the beach to seeing swirling smoke patterns on a drive through the Himalayas, they wondered, observed, and explored.

Through their curiosity, they discovered something they found remarkable: a field of science called *fluid dynamics*—the study of how air, water, and other fluids move and flow.

Today, these girls are scientists, engineers, and problem solvers. They work in exciting careers where they continue to satisfy their curiosity. Their stories—part story, part science—show that curiosity can take you on an adventure anywhere.

Some stories are about a single moment of curiosity and discovery. One girl, determined to shave seconds off her swim time, learned that the position of her body in the water reduced drag, letting her swim faster. Another girl, fascinated by how the wind made the leaves fold, ran an experiment on a hike with her dad. Armed with a fan, she tested how much each a leaf bent based on its size and stiffness.

Other stories follow a longer journey, from that spark of curiosity to becoming a fully-fledged researcher. One girl, noticing her grandmother's struggle with memory loss, took on the challenge of designing tiny fluids labs to help doctors develop new medicines faster. Another girl, in love with both science and sports, discovered how the way she released a cricket ball could make it swing through the air and deceive the batsman.

These stories don't just stop at curiosity—they take you deeper into the science of how things move and flow. Along the way, you'll learn about fluid dynamics alongside these girls. You'll find even more science in the Curious Girls Notes. And you'll see just how far your curiosity can take you.

If you find yourself even more curious as you read, don't worry! When you reach the end of the last story, the epilogue will guide you deeper into the world of scientific research and show you how to follow your own curiosity into the field of fluid dynamics.

The women who wrote these stories come from different cultures and countries. Even so, they all started out just like you—curious, observant, and full of questions.

Now they are scientists, engineers, and experts in fluid dynamics. They are students, researchers, and professors working to solve problems and unlock mysteries. Their work includes everything from large rockets lifting into space down to the small blood cells traveling through our veins. Their stories will show you the joy and challenges that working in science can bring.

As you turn these pages and meet each of us, we hope that you find a little bit of yourself in each story and learn a little about the science we are so passionate about. Who knows? Maybe one day, you'll join us and become the next fluid dynamicist uncovering the secrets of the world that flow around us.

Keep following your curiosity,
THE SOWIF INITIATIVE

windswept thoughts

The Story of a Girl Who Is Curious about Leaves

MRUDHULA BASKARAN

My favorite oceanographer Jacques-Yves Cousteau once said, "When one man, for whatever reason, has the opportunity to lead an extraordinary life, he has no right to keep it to himself."

Here I will share a discovery I made when I was younger, which rendered my life truly extraordinary ever since.

It was Friday, and I had just returned home from a day of school. I was ten.

My homework assignments lay in my school bag. I would get to them later that weekend. Now, I had something very exciting to look forward to. I shook the mud off my shoes at the main door and ran to my room to pack my backpack. My father and I were going backpacking in the forest, and I needed my essentials.

My notebook was where I kept all of my lists. I flipped to the page that had the essentials for this trip:

1. Binoculars
2. Notebook + pencils to draw
3. Polaroid camera
4. Bag to bring back leaves

I glanced around my bedroom; my walls were covered with photos and glass cases of leaves I'd collected on previous trips—maple, oak, ginkgo, birch, and poplar, to name a few.

I adored leaves. The only thing I loved more than their beauty was the way they moved. The way they danced in the wind. How they shifted when the wind blew against them. How they sometimes gently whistled or resiliently hung on to their branches despite the wind threatening to rip them away.

Backpacking was a combination of all things I loved: Walks in the forest. Being outdoors. Making campfires and doing things with my hands. I had always known I learned best by doing, and during backpacking, the world became my canvas and my ideas, the paint.

The next morning, my father drove us to the start of the trail, and we began our hike.

Instantly, he said, "I have a surprise for you."

"What is it?"

"I brought a tent for us to camp outdoors!"

The day was off to a fantastic start. It was sunny, and the forest was quiet. We began walking up the mountain, passing by oaks and poplars, aspen trees whose branches showed new buds emerging, and pines that hadn't lost their needles over the winter.

We stopped in the shade of a maple tree for lunch. The wind that had been calm and well-behaved so far decided to strengthen. The trees near us were bending in the wind together. Their thickets of small leaves were all bending together, and the entire tree seemed to bend as one unit.

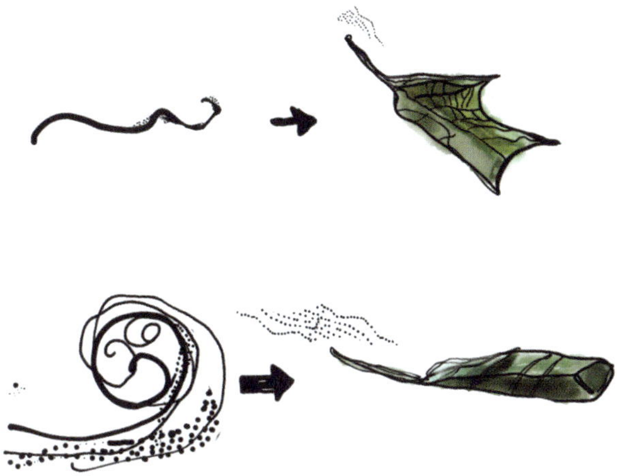

It looked like the behavior of wheat fields I had seen on windy days at a farm.

Looking closely at the maple leaves above me, I observed something completely different from the trees with small leaves. Each individual maple leaf had folded in on itself. Some leaves were merely folded in half while others were completely rolled up like small carpets in furniture shops.

Why is it that some maple leaves are rolled more tightly than others? I must have wondered this aloud because my dad responded to my question with a question.

"Do you think the wind has something to do with it?"

I finished my sandwich in two bites and stood up from the stump where I was seated.

I plucked a maple leaf from the tree and held it in the air above me. When the wind blew very violently, the leaf curled in on itself, and it was difficult to hold on to the small stem.

When the wind was gentler, the leaf bent less. I realized the connection between how hard the wind blew and how much the leaf responded. I thought back to earlier that day when we had just started the hike. The wind had not been so intense, and the leaves of trees did not curl up or move around. They had just lain still and tried to soak up the sun.

I tried to demonstrate this to my dad. "Hey, look!" I yelled, holding up the maple leaf. But the wind was blowing very randomly, and there would be strong gusts. It was hard to show the connection between the wind strength and how much the leaf folded around itself when I could not precisely control the direction or the strength of the wind.

My dad chimed in, "The wind is definitely hard to predict. Can you think of a way to get a more controlled flow of air? Let's keep going and make camp soon. Maybe you'll get some ideas along the way."

I plucked a few maple leaves and stuffed them carefully in my plastic bag. My head teemed with ideas the rest of the way to the summit. How could I control the wind? I could try blowing against the leaves while I held them still. But was there a way to get the wind to blow gently and then more strongly against the leaves exactly when I wanted them to?

The answer appeared momentarily.

The mountaintop lay above the tree line. The sun peered out from behind the clouds radiantly, covering us in heat. The wind seemed to have disappeared. Sweaty from the exertion, my father reached into his bag and pulled out a small, battery-powered fan and began moving it around his face. Suddenly, he stopped and stared at the fan. The idea must have hit us both at the same time, as he turned my way and grinned.

Snatching the fan from my father's outstretched hand, I shouted, "Eureka!"

I grabbed a maple leaf from my bag and held it by its stem with my left hand. I turned on the fan with my right hand, keeping it at its lowest intensity and holding it an arm's length away from the leaf. The leaf bent gently and just began to fold along the stem running through the center of the leaf. When I increased the fan's intensity, the leaf folded and then it began to roll up, just like the maple leaves had done in the heavy wind.

My newfound discovery thrilled me, leaving an aftertaste of invincibility. My test had shown that leaves respond directly to the way the wind blows.

If there is no wind, the leaves are still. When there is wind, they adjust how they respond, whether by lightly bending or fully rolling up.

I thought, *Surely all leaves must respond in the same way.* Earlier, I saw how the trees with small leaves bent as one big group because the branches were so thin. The maple leaves, which were larger, folded in on themselves. All of these leaves were trying to reduce how much of them the wind was trying to blow against.

I instantly began testing my theory on whatever plants I could pull up around the mountain summit. Tall blades of grass bent more in stronger winds from the small hand fan, compared to gentler winds.

I ran to a spot on the summit where a dark green plant with thick leaves was growing.

My dad chuckled and said, "Indian rubber plant. That is a ficus!"

I grabbed a fallen leaf from beside the ficus and held it up to my hand fan. I then watched my moments of scientific joy collapse before my eyes.

The ficus leaf remained still and refused to bend. I increased the intensity of the hand fan and desperately started blowing at the ficus leaf, but it refused to roll up like the well-behaved maple leaf had done. A tear slipped down my sweaty face. My success as a scientist had been so short-lived. My inner Romulus Augustus emerged as I cried, "'Tis the fall of Rome!"

My dad heard my cry of disappointment. "What do you think is different about the ficus? Let's think of it as we make camp."

I grabbed some more fallen ficus leaves dejectedly, and we left the summit.

"Don't feel down about not finding all the answers. Science is bit like that. It takes its own time for the answers to appear!"

I sat before the campfire my dad built to brighten my mood, replaying the day in my head. Why had the ficus refused to bend? I fiddled with a blade of grass, bending it back and forth, until it suddenly snapped. That's when it hit me: Different plants are tougher to bend than others. I grabbed another blade, this time a bit thicker, and bent it this way and that, deep in thought.

Minutes later, I looked down at the grass in my hand, noticing how that blade of grass had remained resilient to my forceful bending, while the other blade was instantly torn up. I had just been playfully bending these blades of grass, not really putting too much of my strength into it. And yet one remained so much more robust than the other.

A realization hit me: Could different plants require different strengths to force them to bend? Could a maple leaf be easier to bend than a ficus, just from differences in the inherent nature of the plants? I brought out a ficus from my bag. I had to push very strongly against the leaf to get it to bend. The maple leaf required almost no force and bent instantly.

CURIOUS GIRL NOTE—Why do only some leaves roll up in the wind?

- They are trying to reduce how much of them the wind tries to blow against.
- Different plants require different wind strengths to force them to bend.
- Maple leaves bend easily, but ficus leaves need a more forceful wind to bend them.

The wind slowly picked up. Earlier we had heard a soothing whistling of the air passing through leaves. Now, we heard howls and shrieks.

I held the ficus stem out in my hand and watched as the wind pelted the leaf. Finally, when the wind hit the leaf forcefully enough, I noticed a small bending of the ficus leaf. It slightly began to fold on itself just as the maple had done in gentle winds. A soothing calm washed over me as I clasped the leaf with both hands.

I retreated into my tent and pondered over my discoveries.

CURIOUS GIRL NOTE—Discoveries of the day:

- Plants bend more in stronger winds than in gentler winds. Some plants even roll up!
- Plants behave differently to winds of the same intensity. A ficus leaf will bend less than a maple leaf in the same wind.

"Was the discovery about the ficus reasonable?" I asked.

My dad chimed in, "Well, a ficus on top of a mountain has to deal with much heavier winds. It has to be hardier than a leaf on the side of the mountain. It makes a lot of sense that it takes more force to bend a ficus than a maple leaf."

A ficus on top of the mountain did have to deal with harder weather conditions than a tree below the summit. The summit was mostly filled with hardy shrubs—very rarely trees. The fervor of discovery, which had temporarily been hushed when I had first picked up the ficus leaf, returned. The cries of the wind gave voice to my exhilaration.

Different plants survive in different ways, but they also have some things in common. Most plants are flexible, even if some bend more easily than others. A ficus is flexible but not as flexible as grass or a maple leaf. And depending on how hard the wind blows, plants will bend different amounts.

I turned on the fan and watched the maple leaves roll up. I held my hand outside the tent and watched the ficus bend in the merciless wind. I stepped inside and closed the door. Retreating calmly into my sleeping bag, I fell asleep in the warmth of my own satiated curiosity.

super sensor

The Story of a Girl Curious about Seal Whiskers

BANAFSHEH SEYED-AGHAZADEH

As I walked toward the meeting room, I felt super excited to meet my team of graduate students and discuss what I thought could turn into one of the coolest projects in the lab. Being a professor had its perks—especially leading a research lab full of brilliant students who were always ready for the next big idea. But as I opened the door, I saw everyone sitting around the table with anxious faces, clearly wondering what this sudden, out-of-schedule meeting was about.

"Do you want to work on a whisker project?" I could hear this question in my head and couldn't help but laugh at the thought. The question was followed by bewildered looks. They probably thought I was sleep-deprived from my newborn or under too much pressure from work. They looked puzzled at each other and finally gave me polite, awkward smiles.

"Sure . . . but did you say whisker? What about whiskers?" one brave student finally asked, smiling but clearly still puzzled.

Now it was time for me to sell the project. "Well, not any whisker, but super whiskers!" More silence in the room. "Okay, not exactly super whiskers, but harbor seals' whiskers." Still, no response. I continued, "You know, harbor seals have a unique capability to use their whiskers to sense their surroundings in water, which, if you think about this, ties into fluid dynamics—the general theme of our research lab!" I gradually saw their interest pique.

"So what do you want us to test exactly?" they asked.

"I want us to figure out how this sensing in the fluid environment works," I replied.

"But why is this topic important? Shouldn't a biology major be looking into this problem? I can talk to my biology professor for help," one student suggested.

"Hold your horses!" I said with a smile. "The reason this topic is so important is that we can use how whiskers sense things to inspire new underwater sensors and tools that help avoid collisions and track targets. Think about how a seal uses its whiskers to feel around and avoid bumping into things in the water.

"But that's not all—seals also use their whiskers to follow the movements of fish they're hunting. This can help underwater robots avoid obstacles and follow moving targets."

CURIOUS GIRL NOTE—Seals use their whiskers to sense the environment and hunt for fish:

- Whisker shape: flat, wavy, up-and-down bends.
- The shape and bending of the whiskers helps the seal detect the flow.
- Can we design underwater sensors and tools based on how seal whiskers "feel" the flow?
- Can we avoid obstacles and follow moving targets?

I turned to the student who mentioned their biology professor. "While this is our project, and I'm confident we have the skills as mechanical engineers to tackle it, working together with biology experts can definitely enhance our understanding."

Catching my enthusiasm, they left the room buzzing with questions and interesting ideas.

Many months later, after countless hours of studying the biology of marine animals, discussing various design, modeling, and manufacturing techniques, my team gathered in that same room and presented their findings with palpable excitement.

"When a fish swims nearby, it leaves behind a trail of swirling water, like footprints in the sand."

"As seals glide through these swirling trails, their whiskers dance among the whirls, causing their cheeks to twitch and guiding them to the fish that left the footprint."

It was their turn to surprise me. "Do you know what an actual whisker for the seal looks like?"

Like a magician pulling a rabbit from a hat, they pulled a replica of a tiny whisker from a sack.

It looked strange! Unlike a typical whisker with a round, uniform shape, this whisker was flat and had wavy, up-and-down bends along its length.

"Where did you find this?" one asked.

"We found 3D-scanned dimensions of an actual whisker taken from a seal at an aquarium. We used these dimensions to create a 3D model in computer-aided design software. Thanks to the 3D printer in our lab, we were able to print a scaled model of the whisker, capturing every tiny detail you see here."

This whisker project was now gaining momentum!

With the 3D-printed replica of a seal whisker in our possession, we felt like the cool kids of our department! We seized every opportunity to showcase this oddly shaped, crooked whisker replica, surprising audiences with the intriguing project we were working on. We held it dearly, almost like a trophy, during meetings and lab visits by others.

Our aim was to better comprehend the intricate sensing mechanism of the whisker and potentially design futuristic underwater sensory systems inspired by it. We needed to understand how the water and the flexible whisker work together in a back-and-forth way to create this sensing ability.

Luckily, this is exactly what my lab studies—we focus on how moving water interacts with different shapes, like whiskers, and they get engaged in a *dynamic conversation*.

When I say there's a dynamic conversation between the water and the whisker, I mean that the water and the whisker are constantly affecting each other. As the water moves, it pushes on the whisker, and the whisker bends and moves in response. This back-and-forth helps the whisker sense what's happening around it.

In my lab, we study how things like water flow and different shapes, like whiskers, interact with each other. This is what we call *fluid-structure interactions*—how liquids like water and solid things like whiskers affect each other when they come into contact. However, this was the first time my team was

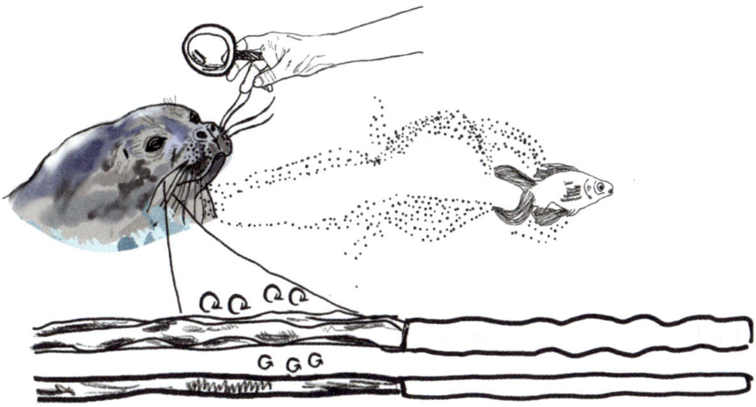

working on a topic with strong ties to marine biology, making some aspects of the project both daunting and intriguing.

Our next step was to understand what about the whisker made it such a super sensor for the harbor seal. We 3D-modeled and printed different versions of these tiny whiskers. By placing the whisker model in the water tunnel, which was like a wind tunnel but filled with water, we created our own little swimming seal.

Our first aha moment was when we saw the whisker start to vibrate as it responded to our artificially created fish footprints.

This vibration started from the tip of the whisker and travels down its length to where it attached to the seal—or, basically, the seal's nervous system. That was likely how the seal sensed the footprints!

Research took off then as we created different types of fish footprints to induce different types of vibrations in the whiskers. By mimicking the wakes created by swimming fish, we could better understand the complex flow dynamics that influence the whisker's vibration response.

In our experiments, we added tiny particles to the water in the tunnel and shone a powerful light on them to track the particles' movement. This let us measure the water's speed and direction, which was invisible before—like putting on magical glasses that reveal the hidden world of water flow! We also used other methods to visualize how water moves, such as adding a harmless dye or creating tiny hydrogen bubbles through a process called *electrolysis*. All of these experimental techniques helped us measure how the harbor seal's whisker interacts and follows the wakes of the fish.

CURIOUS GIRL NOTE—How can we figure out why whiskers are super sensors?

- With 3D modeling and printing based on real whiskers;
- Put the "whisker" into a water tunnel (like an underwater treadmill);
- Add tiny particles to the water and shine a light to track how the water moves; and
- Test different "fish" flows in front of the whisker to see how the flow changes around the whisker.

Working with other researchers, we discovered that the special shape of the harbor seal's whisker helps it detect movements in the water, even from far away. These whiskers are so sensitive that seals can tell the difference between a fish that's swimming fast and one that's just leisurely gliding through the water, not realizing a predator is nearby.

This exciting discovery might help us create new sensors that can avoid collisions, just like a seal's whiskers do. Our research not only teaches us more about how marine animals like seals live but it can also lead to important advances in technology. It's amazing to think that by studying these playful seals, we might change the way we design new tools and devices!

Months after that out-of-schedule meeting, I found myself sitting at the back of a packed conference room. My team was at the front, proudly presenting our "whisker." That odd little replica, once a punchline to my sleep-deprived-professor moment, now sat like a trophy in the spotlight.

One slide after another, they unraveled the story—how the seal's whisker dances through water trails, how its vibrations hold clues to unseen movements, and how our experiments turned this natural wonder into inspiration for future underwater sensors.

I couldn't help but smile. I remembered their faces that day—puzzled stares, cautious nods, and that brave, *"Whiskers? What about whiskers?"*

Now here we were, watching a quirky idea grow into something remarkable.

As the presentation ended and applause filled the room, I leaned back and thought, *Sometimes, all it takes to unlock big discoveries is a small question, a lot of curiosity, and a team willing to follow an unusual idea—no matter how strange it sounds at first.*

And honestly? That's what makes this job pretty cool!

from mosquito whirs to aerodynamic whirls

The Story of a Girl Curious about Nature's Smallest Pilots

SWATHI KRISHNA

It was the beginning of summer in Southern India. The end-of-year exams were finally over. The holidays had begun!

The smell of raw mangoes wafted through the air as I walked home from school. The afternoon felt quiet and lazy, and I was tired from the exam. A nap seemed like the perfect idea.

The moment I got home, I jumped into my pajamas and flopped onto the bed. As soon as my head hit the pillow, the rhythmic creak of the ceiling fan began to lull me to sleep. My eyes grew heavy, and I was just about to drift off when—*whirrrrrrrr*! A mosquito's high-pitched whine shattered the silence and rudely awoke me from my slumber.

Summers in tropical regions like South India create a perfect playground for insects, especially mosquitoes.

Ah! They are back to torment me! I thought. Still lying down, I waved my arms frantically to get rid of the pesky little brat, but it wouldn't leave me alone. Grandma's voice popped into my head: "*You can't just wave a mosquito away—you've got to catch it or swat it!*" I sat up grumbling, determined to face my tiny, winged enemy.

Waiting patiently, I devised my strategy. I peered around the room searching for the source of the whirring sound—just like a Quidditch seeker scanning for the golden snitch. Then I spotted it! The tiny mosquito zipped around the room, circling the mirror, vanishing and reappearing faster than I could keep up with.

It didn't seem interested in me now that I was awake. *How do they fly so fast—and in every direction like that?* I wondered. I kept my eyes locked on it as it finally took a break, resting on the windowsill near my bed. It seemed completely unaware of me.

Slowly, inch by inch, I crept closer, careful not to make any sudden moves that would give away my intentions. My hand hovered, ready to strike. In one swift motion, I lunged forward—but it was gone! The mosquito had slipped away again. I looked around, but the tiny snitch was nowhere to be seen. *How did it take off so quickly?* I sat back, stunned by its lightning-fast getaway.

Over that summer and many others, I realized that it wasn't just mosquitoes that managed to escape me. There were bees, flies, and moths—each as elusive as the last. Then there were the larger, scarier ones like flying cockroaches, which I made sure to keep far away from!

But the most annoying ones were the insects that seemed to hover right in front of my face, taunting me. I couldn't help but wonder, *How do they manage to stay in the air, in the same spot, without falling?*

Years later, as a student of aerodynamics, I finally found the answers to all my childhood questions. Back then, I'd thought that the mosquito's high-pitched whine was its way of singing (very badly, I might add). But I learned that the sound we hear is actually from their wings beating around eight hundred times per second!

This rapid wing motion not only creates their unmistakable hum but also lets insects move in incredible ways. They can hover in one spot, dart sideways, fly backward, accelerate faster than the blink of an eye, and even travel long distances. Over millions of years, these tiny creatures have become masters of flight.

As time passed, I started to draw comparisons between the mosquito's effortless dance and the majestic human-built airplanes I saw overhead. My earliest memory of this obsession dates back to when my dad took me to an airshow. I was ten years old, standing in awe with a stick of candy floss clutched in my small hands, as he excitedly pointed out the different planes decorating the sky.

"Look at that helicopter!" he said. "It's hovering perfectly in one spot. How do you think it's doing that?"

The loud whirring blades reminded me of something very tiny. I squinted curiously and asked, "Is it flying like a mosquito?"

Dad chuckled. "Not quite. Helicopters spin their wings in circles, like a fan. But keep watching—see that fighter jet?"

Before I could reply, the jet flipped midair, zooming past us and performing somersaults in the air.

"Whoa!" I gasped. "That's so fast! Why don't all planes do that?" I jumped up and down.

"Every plane has its own strengths and weaknesses. I don't think I can name one aircraft that can perform all those acrobatics your favorite mosquitoes do!" he teased.

I was mesmerized by the spectacle as tiny planes wove through small gaps, and far in the distance, a massive passenger aircraft glided smoothly through the blue sky, looking like a slightly big insect itself.

Why can't all these big airplanes fly like mosquitoes? I thought.

On the way home, still reeling from the day's excitement, I wanted to fly right then and there! But stuck inside a bus, all I could do was put my hand out of the window (with Dad's permission, of course), holding it flat like an airplane wing.

"Tilt your hand up a little," Dad said, noticing what I was doing.

I followed his advice, and suddenly my hand was pushed upward by the rushing air. "Whoa, it feels like my hand is flying!" I exclaimed.

That innocent moment sparked a question I would chase for years: What really makes something fly? I pored over books, toured factories where they designed and built planes, and spent summers at science camps. Eventually, I found myself at university studying and researching about *aerodynamics*—the science of how air moves over objects.

The more I explored, the clearer it became. Whether it's a whirring mosquito or a giant airplane, the wings are the real heroes of flying! How the air flows around the wings determines how well either of them can take to the skies!

Turns out air doesn't flow in the same way around airplanes and insect wings.

Let me explain with something simpler—big passenger airplanes! Ever notice how their wings are stuck firmly to their sides? They don't flap or wiggle like an insect's wings, which is why we call them fixed-wing aircraft.

Air is a fluid, much like water or even honey. All fluids have some "stickiness" that prevents them from slipping on a surface. While air is not as sticky as honey, it has just enough stickiness to cling to the surface as it flows over the wing. Also, here's a fun fact—air is a tiny bit lazy, like us! It doesn't want to work any harder by taking a tricky or a long path if it doesn't have to. So when an airplane's long, straight wings slice through the air, the air follows the surface of the wing snugly, like a cozy blanket.

Now, as the plane speeds up, the shape of the wing causes the blanket of air to slide toward the rear edge of the wing and gets pushed downward and away from the wing. As a reaction to this downward movement, there is a strong pushback from the air below the wing. This *pushback* is the lift force that helps the plane rise to the sky.

Here's the catch: Airplanes can only stay in the air if they keep moving forward. Their wings need speed to keep generating lift. If a plane slows down too much, it can't push enough air down, and—uh-oh!—it starts to fall. Unlike mosquitoes, airplanes can't just stay in the air in one spot. They always need to keep moving forward, deflecting air downward to stay up.

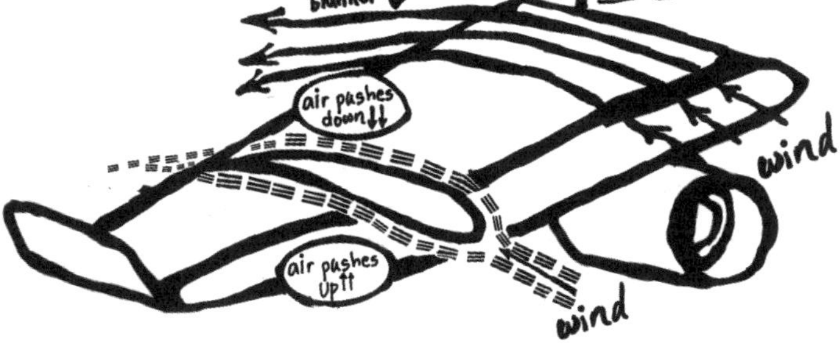

To see how this works, scientists test wings in wind tunnels—giant tubes with powerful fans that blow air. Inside, they place models of planes to see how air moves around the wings and measure how much lift is being created. Based on their measurements, they tweak the shape of the wings to improve their performance.

When I was at university, I got to work in one of the largest wind tunnels in India. Watching the air flow around wings and seeing lift in action sparked a fascination with the hands-on, experimental way of visualizing the invisible world of air movement.

This experience got me thinking. We understand how airplanes fly, but they still can't do everything that insects do. What makes insects so different? How do their tiny wings let them perform such complicated aerial feats? These questions stayed with me until my PhD research, where I finally had the chance to dive into the mechanics of insect flight.

The answer lies in the way the insects move their wings. In contrast to our earlier example of a fixed-wing aircraft, insects flap their wings with astonishing vigor, moving them in not one, not two, but three directions at the same time!

Want to try an experiment to understand how insects flap their wings? Let's give it a shot!

Step 1: Extend one arm straight out to the side so it's parallel to the floor, with your palm facing forward. This is your base position.
Step 2: Now move the entire arm to your front and then all the way to your back gently. This back-and-forth movement mimics the main *stroke* motion of an insect wing.
Step 3: Return your arm to the base position. Now rotate your palm to the floor and then toward the ceiling like flipping a pancake. This second motion mimics the *pitch* of the wing.

Step 4: Finally, bring your arm back to the base position and move it up and down slightly, as if it's making small adjustments. This is similar to the *elevation* motion of an insect's wing.

Step 5: Now try doing all three motions—stroke, rotation, and elevation—at the same time.

If you do this correctly, your hand will trace a sideways figure-eight pattern in the air. Do this with both arms and you will be mimicking the complex wing movements of insects. But here's the real kicker: Insects perform these motions hundreds of times per second! Isn't it impressive how their tiny bodies can achieve such incredible speed and coordination?

This complex movement is what we call *flapping wings*, which is very different from the stationary, rigid, fixed wings of an airplane or the spinning rotary wings of a helicopter.

If their wing movements are so complicated, how exactly do insects produce lift that is necessary to keep their tiny bodies in air?

Let's consult one of the experts!

Imagine we're out in a sunny park on a warm, spring afternoon. As we sit by a patch of flowers, a busy bumblebee buzzes up to us.

"Hello there!" she hums. "Word amongst the flowers is that you are looking for a flying expert? Look no further. I'm your bee!" she says proudly, adjusting her tiny wings. "Let me show you my most important trick—it's how I hover in one spot while carrying loads of nectar. Watch carefully!"

With that, she traces the first half of the figure-eight motion with one of her wings, and something incredible happens: A swirling pocket of air forms above it, like a mini tornado.

"See this spinning air here? It's called a vortex. Doesn't it look like the whirlpool you see when you pull the plug in a bathtub?" She chuckles excitedly and continues, "The center of this vortex has super low pressure, like the calm eye of a storm. Meanwhile, the air beneath my wing just flows gently by, applying a higher pressure that pushes me upward. That's how I generate lift!"

CURIOUS GIRL NOTE—How does a bee fly?

- Vortex: spinning air (like the whirlpool in the bathwater when you pull the plug).
- Center of the vortex has low pressure.
- A vortex above a wing sucks it upward (air below the wing flows by without a vortex).
- A vortex dies away over time.
- Insects have to generate a vortex with each stroke, many times a second, to be able to fly.

She pauses briefly as her wings reach the end of their stroke. "But here's the tricky part," she adds, leaning in as if sharing a secret. "These vortices don't stick around! They vanish every time I switch the direction of my wing stroke. So I have to flap my wings and create a fresh vortex on each wing constantly—hundreds of times per second! It's hard work, but it's what keeps me airborne. Neat, huh?" And with a proud little flourish of her wings, she finishes her demonstration.

As she prepares to fly away, she quips, "You can take some pictures of my vortex if you'd like." With that, she tosses some sparkly pollen into the air and flies through it. The pollen is drawn into the vortex, making it glow brighter and more visible!

Luckily, I'm ready with my camera. As an experimental aerodynamicist, I take high-speed photographs all the time, like the burst mode on a phone camera. I get a series of pictures, and when I play back the images, it's like flipping through a picture book showing how the vortex forms, grows big, and vanishes in slow motion. It's not just science—it's beautiful!

Different insects flap their wings at different speeds, but they all rely on the vortex to stay in the air. Pretty impressive for such fragile-looking creatures, right?

There are many ways to study insect flight. But between you and me, as much as I love learning about flying insects, I'm not exactly a fan of catching live ones. They're too wiggly, and I wouldn't want to hurt them!

That's why in my lab, I use a robot that flaps its wings just like an insect. We sprinkle sparkly dust into the air, watch it swirl into tiny tornadoes, and use special sensors to measure the lift these whirls create. Robots let us control wing movements completely, letting us explore possibilities beyond what insects can do naturally.

In one experiment, my team and I discovered that with just a few tweaks to the wing motion, we could double the lift without using any extra power. This discovery could help us design flying machines that outperform insects in some ways!

The more we learn from these tiny acrobats, the smarter and more adaptable our flying technologies become. These insights are helping us create better aircraft with fewer weaknesses and more strengths.

So the next time you see a bee or a fly or a mosquito whirring around you, think about the magic of thousands of aerodynamic whirls keeping it aloft. These miniature acrobats are nature's most remarkable fliers—and they are helping us reimagine the future of flight.

swimming faster and smarter

The Story of a Girl Who Dreams of Going to the Olympics

ELIZABETH GREGORIO

One long whistle cuts through the noise on the pool deck. I step onto the diving block, ready to swim the 100-meter freestyle. I place my feet in the starting position and shake out my nerves to focus on the race.

The official says, “Take your mark.”

I bend down, gripping the edge of the diving block and looking out to the pool.

When the buzzer sounds, I launch myself into the air. I glide for a moment before slicing into the water. When I feel myself slowing down, I start my dolphin kicks, and when I need air, I start swimming. I move my arms and legs as fast as I can. But it feels like I'm fighting the water with every stroke.

I keep pushing myself for all four lengths of the pool. I slam my hand into the wall at the finish and look up to the scoreboard to check my time. It's okay, but it's the same as last meet's. I was really hoping to go faster today.

I climb out of the pool and go over to see my coach. Coach Anne pats me on the shoulder and says, "That was a good swim. Go drink some water and rest."

"Okay," I say even though I don't mean it. I've been getting the same time for a couple of months, and it's so frustrating. I want to get faster!

I rejoin my teammates to watch the rest of the meet and chat about which races we're most excited to watch in the Olympics next week.

When the meet ends, I wave goodbye to Coach Anne and head out to find my parents. Maybe watching the Olympics will inspire me to beat my best time at the next meet.

I love watching the Olympics. It's the one time every four years when I can watch swimming on TV. Plus, the atmosphere is so electric, you can feel it through the screen. This year is especially exciting—the swimmers are breaking world records in just about every race! The announcers keep saying it's because of their new swimsuits.

The suits cover almost the whole swimmer. It looks like they're wearing a tank top and leggings, almost like what a superhero would wear to fight crime. But the swimmers are wearing it to go faster than anyone thought was possible.

I could definitely beat my 100 freestyle time with one of those.

Of course, Coach Anne would say I should just keep practicing. So I try to remember that as I open the door to the pool for my first practice after the meet. I see my coach in the front office and walk in.

"Hey, Coach Anne!"

"Good afternoon, Elizabeth. Are you ready for practice today?"

"Yeah, I am! I really want to get my 100 freestyle faster for the next meet. Have you been watching the Olympics? The new swimsuits are working like magic to make everyone faster."

Coach Anne laughs and replies, "It's not magic. Those swimmers worked really hard to get to the Olympics. But it does seem like they're getting a boost with those suits."

I smile; I knew she'd say that. "Yeah, of course they've worked hard, but it's cool that the scientists designed the suits to be like sharkskin. I didn't think scientists would be interested in swimming."

"That is interesting. I wonder if you could find any other things scientists have done to help swimmers. But first you have to go practice if you ever want to be fast enough to go to the Olympics yourself."

"Okay, Coach Anne. I'll see you on the deck."

I've been swimming for Coach Anne for over a year now. I joined the team about a month after I started sixth grade. She's a tough coach, and I've heard that she was an even tougher chemistry teacher before she retired. If she thinks that there are other things that scientists have done to help swimmers, I bet I'll find something cool.

I love being a swimmer.

When I was little, I wanted to be a swimmer so much that I begged my parents to put me into swim lessons. I made it through all of the levels pretty quickly and then my teacher told my mom to put me on a swim team.

So here I am. But it's not just about swimming; I love the water. The smell of chlorine instantly makes me feel at home. I love the feeling of being engulfed by the water when I jump into a pool or plunge into the ocean. And I love pushing myself to work as hard as I can to get as fast as I can—even when it's frustrating to not beat my best time in every race.

Being a swimmer has made it even more exciting to watch the Olympics too. Now I understand some of what it takes to get there, and I hope I can make it there someday too. I want to compete with all the fastest swimmers in the world. All of us doing our favorite thing. Showing off the rewards of all our years of training.

For now, I have to focus on doing that training.

I step out onto the pool deck in my swimsuit and stand with my teammates. Coach Anne stands in front of the board writing out the warm-up sets for each of the age groups.

She turns around and says, "I've got a good practice for all of you today! I want you to remember to stay in streamline for as long as you can every time you push off the wall. Let's make the most of those underwaters, which means no breathing until you've done two strokes. Okay, what are you waiting for? Let's get going!"

We move to our lanes, put our caps and goggles on, and start jumping into the water.

When it's my turn to jump in, I only hesitate for a second at the thought of how cold the pool will be. As soon as I'm in, I start swimming to try to keep my body warm, which is probably why we call it a warm-up.

Our warm-up set is long, so I let my mind wander. I think about what Coach Anne said about the underwaters, the part after you push off the wall and before you start doing strokes again. I try to do some dolphin kicks when I push off from my flip turns, but it's hard to hold my breath for long enough to do that *and* take two strokes before I breathe again. I wonder how the older kids seem to do it so easily.

Then I wonder what number lap I'm on. I notice my friends stopped at the wall ahead of me. Luckily, they are much better at counting. I always get distracted, so I'm not allowed to go first in our lane.

It's a hard practice. That's always what Coach Anne means by "I've got a good practice for you." Today we had to swim more yards than I ever have before. All the sets were focused on improving our underwaters, and holding my breath gets harder and harder as practice goes on.

But I notice that doing dolphin kicks off of the wall doesn't get harder. It actually feels less like I'm fighting against the water than when I'm swimming at the surface. I wish I could make my surface swimming feel just as smooth.

For the rest of the practice, I keep reminding myself that hard practices make for fast swimmers. But that doesn't change how tired I am when we all finally get to climb out of the pool.

When I'm ready to leave, I stop by Coach Anne's office. "Bye, Coach Anne. I'll see you tomorrow!"

"Bye, Elizabeth. I'm looking forward to hearing about other examples you find of scientists helping swimmers."

"Me too! I almost forgot about that after such a hard practice."

She laughs and waves as I walk out to meet my mom.

After dinner, I have an hour before I can watch the Olympic swimming on TV, so I head to the computer to do some research.

I start by looking up some articles about the swimsuits. The scientists they interviewed talked about how they were able to reduce the drag of the swimmer's skin by making a fabric like sharkskin.

The mention of drag reminds me of when the older kids wear five swimsuits on top of each other at practice. They say it increases their drag to make it harder to swim in practice. Apparently, that makes it easier to swim fast in a meet when you're only wearing one swimsuit.

Hearing the scientists mention it, too, makes me think that drag could be important. So I look up the definition:

> *Drag* is a force that opposes motion of an object moving through a fluid.

Which makes a lot of sense in swimming. Drag must be what we have to overcome to swim fast! Maybe it's drag that I was feeling during my 100 freestyle race when it felt like I was fighting against the water.

I write down some notes to talk to Coach Anne about tomorrow and check again when the Olympic swimming is going to start.

I get to practice early the next day. When I walk into the office with my notebook, Coach Anne smiles.

"So what did you find out?"

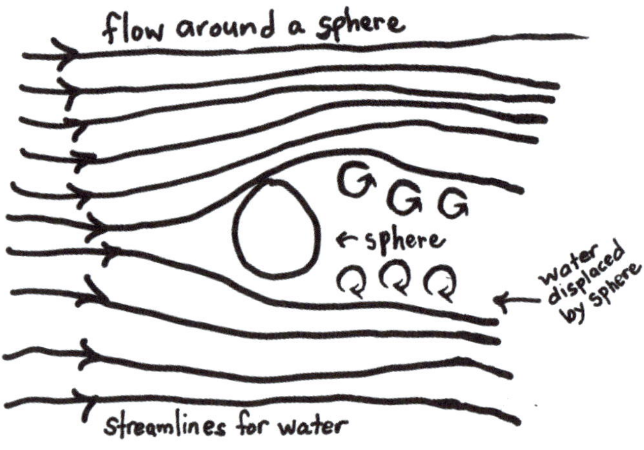

"I learned what drag really is. Look, I drew a diagram." I show her the page in my notebook. "All of these lines are called streamlines. They show where the water goes when you move the sphere. The water that's far away just goes on as it always has, but the water right next to the sphere moves a lot. All of the water that the ball has to push out of the way creates the drag force."

I look up at her and demonstrate a streamline with my hands over my head. "That's why I have to keep my body tight for streamline. I don't want to disturb more water than I have to."

CURIOUS GIRL NOTE—How can I reduce drag so that I can swim faster and go to the Olympics?

- *Drag:* A force that opposes the motion of an object as it moves through a fluid.
- Streamlines show how the fluid moves.
- Example of drag: A ball in flowing water. The ball has to "push" the water around it, which creates the drag force.
- To be more streamlined, keep my swimming form tight to minimize disturbing the water!

She nods her head. "That's exactly right. Do you think you could expand that concept to when you're not in streamline?"

"I've been trying to think about that, because sometimes it feels like I'm fighting against the water. That could mean I'm creating more drag than I should, but I thought to go fast you need big kicks and strokes?"

She nods, considering my question. "Sure, but you want to make sure that you only move water that will help you go forward."

"I guess that means if I'm swimming on the surface of the water, I don't want to move too much deep water."

She smiles. "Yes, you can think of your body like a seesaw. You want it to stay as horizontal as possible. Things like picking up your head too much when you breathe can sink your legs deeper than they should be."

"Hmm, that makes sense, but I think I need to try it in the pool."

"Why don't you try it out during practice today?"

"Okay, sure."

I go last in my lane during warm-up so I can try a different technique on each lap.

First, I swim holding my head above the water the whole time. I notice my feet sinking and realize I have to kick a lot to stay afloat.

Next, I swim normally but pick up my head completely out of the water for each breath. I feel my body seesawing from horizontal to almost vertical.

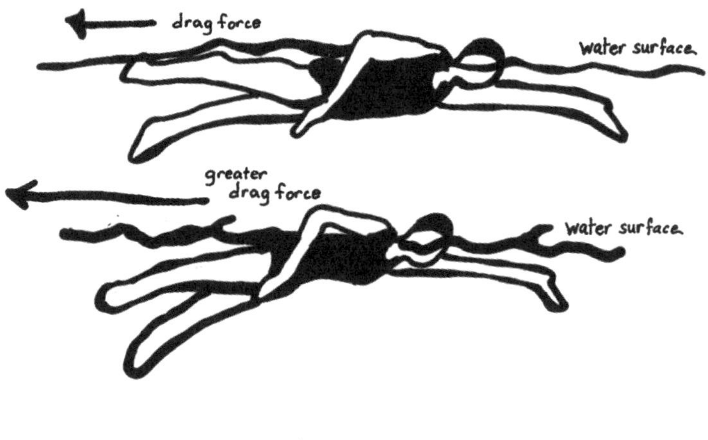

When I swim a lap with my normal freestyle stroke, it does feel smoother. But I notice that I am still seesawing a little, and each breath pushes me backward. It seems that even picking up your head a little to breathe causes your feet to sink.

So I try rotating my head with the rest of my body so that only half my mouth and one eye is above the water when I breathe. This time my body is able to stay horizontal the whole time! I swim the rest of practice like that, trying to get used to a new technique.

On my way out of the pool, Coach Anne stops me. "You really figured out the best way to breathe in freestyle. I bet you'll beat your best time in the next meet now."

A smile spreads over my face, "Thanks, Coach Anne. I hope so!"

One long whistle cuts through the noise on the pool deck, and I step onto the diving block. I'm excited to swim the 100 freestyle again now that I've been practicing the new technique for about a month. I place my feet and focus on the race.

The official says, "Take your mark," and I bend down to grip the edge of the diving block.

The buzzer sounds, and I launch myself into the air. When I pierce through the surface, I make sure to dolphin-kick as far as I can and take two strokes before my first breath.

I'm still pushing myself to swim as fast as I can, but this time I only rotate half my head out of the water to breathe. I feel myself traveling farther with each stroke, and it seems like I'm getting to the walls for each turn faster than I have before.

I keep pushing myself for all four lengths of the pool and I slam my hand into the wall at the finish. When I look up to the scoreboard, I see that I've beaten my time by ten seconds! That's almost a lifetime for a swimmer.

I'm beaming as I walk over to the coaches' stand and get a high five from Coach Anne.

After the meet on the way to the car, I tell my mom about how the new technique I figured out helped me beat my time.

My mom smiles and says, "Wow, it's really cool you could use science to figure this out!"

"Yeah, I had never thought about scientists helping swimmers get faster before. But it makes sense since we're trying to overcome the drag force. It's crazy to think I was actually able to use that to beat my time! It would be really cool to be a swimming scientist someday. But I want to be a really fast swimmer first."

"Well, that sounds like a pretty fun backup plan!"

For the whole ride home, I stare out the window, daydreaming about what it would be like to go to the Olympics over and over again as a swimming scientist who helps swimmers be the best they can be.

moving faster than sound!

The Story of a Girl Who Is Curious about Space Launchers and (Fast) Aircraft

ANNE-MARIE SCHREYER

Lying in the meadow during a summer vacation, watching the clouds and aircraft above me, I wondered where they went and dreamt myself away on adventures, discovering new species and archeological sites in jungles far away. I was yearning to add to the available knowledge and to understand how and why things work. How did that aircraft stay up there in the sky?

I have always been fascinated by all things flying and by rocket launches in the news—the countdown to ignition, the fire, the forces and energy necessary to move those gigantic rockets into space, where they would release satellites or probes to explore planets and stars.

At the same time, I loved nature and the beautiful countryside in southern Germany where I grew up: the hills and forests, rocks and caves, the rivers and lush meadow orchards. It hurt when I imagined how the air and rivers would be polluted and how the orchards and forests would be replaced by airports, roads, and factory buildings.

I was torn between wanting to be a part of developing fascinating new technologies that would fly into the sky and beyond, but also wanting to save the environment and to have a positive impact on society.

Keeping my options open, I searched for an education that would give me the freedom to pursue different career options and choose later. I ultimately chose to aim for a degree in aerospace engineering to get deep background knowledge usable for a broad range of applications, from rocket engines to wind turbines for green energy.

After a while, I realized that the scientific field of fluid dynamics connected many of the applications that interested me. Fluid dynamics was the field where I could study how air allows aircraft and rockets to fly and wind turbines to generate energy. If I went into this area of study, I would not just learn why aircraft and rockets work the way they do, I could also contribute to developing new technologies that create less pollution. I could follow my passion and my conscience.

Before I continue to tell you my story, I need to share with you some of the science about how complicated airflows are when we are designing air and space vehicles.

Air and space vehicles are often flying very fast—even faster than the speed of sound, which is called *supersonic*. Imagine moving so quickly that the air doesn't have time to get out of your way. Rockets need these speeds to escape Earth's gravity, but even modern passenger planes fly so fast that parts of the air over their wings can become supersonic.

Normally, air moves aside gradually when something passes through it—like a car driving on the highway. But at supersonic speeds, the air cannot move out of the way in time. Instead, the pressure waves pile up into a shock wave, creating sudden, strong changes in the air's behavior.

When shock waves form along the surface of an aircraft wing or a space launcher, the flow separates from the surface so that it no longer smoothly follows the surface. It is pushed away from the surface by regions of air that flow slower or even backward. To the outer flow, the wing then seems bigger, and more energy (and fuel) is needed to continue flying.

These separated flow regions create another challenge because they don't sit still. They will dance around the surface of the wing and space launcher, creating a constantly changing pressure field that leads to vibrations in the structure.

While wings and space launchers are made of strong materials, they need to be light, too, and thus are a little bit flexible. The moving, separated flow regions apply varying pressure to the flexible structures, like fingers strumming the string of a guitar. And, like strings of a guitar, the wing's surface material may become brittle after a while from the constant motion, and the material—and therefore the wing—will become damaged.

CURIOUS GIRL NOTE—Supersonic: What does it mean to go faster than the speed of sound?

- Air can't move fast enough when something travels at supersonic speeds → pressure waves build up into a shock wave, sudden changes in how air behaves around the wing.
- Flow separation from the surface is "pushed" away → can make an airplane wing seem bigger! Plus, the separated flow "dances" around, causing vibrations, like strumming a guitar string!
- How do we make safer aircraft to handle these vibrations? Flow control.

Obviously, we want our aircraft to be safe, so we have to make sure that such vibrational damage does not happen. We currently have two options in this important research area:

- Build the wing or vehicle so it is less flexible and thus more stable.
- Modify and control the flow around the vehicle to get rid of the separated regions, or at least stop them from moving around.

Option 1 makes the aircraft heavier and requires more fuel to fly the same distance (and increases carbon emissions into the atmosphere). For space launchers, heavier rockets also need more fuel. To transport more fuel, they can only launch smaller and lighter satellites, making the overall mission more expensive.

Option 2 allows us to build even lighter and, of course safe, aircraft with lower fuel consumption. We can launch satellite systems into space with fewer rocket launches.

Remember when I said that I wanted to contribute to exciting aerospace technology and also reduce the technology's impact on the environment? When I learned about Option 2, I knew that this type of flow control is what I needed to investigate as a researcher.

Reading through scientific papers to find a starting point for my research, I found that one good way to reduce flow separation is to create vortices (plural for vortex) near the surface of the vehicle or wing and get them to help us out.

A vortex is a region in the flow where the air particles follow a whirling or rotating motion, like water flowing out the drain of a water basin. These vortices then bring energy to the air close to the wall, and thus make it easier for the flow to resist separation.

While different kinds of vortices occur in flows all the time, I learned that we can mechanically create them with solid vortex generators (small devices in the shape of small arrows or ramps). If we install these vortex generators onto the surface ahead of the separation, we can control the flow separation and reduce the moving pressure that changes along the wing.

This works well if you fly in a straight line at the same altitude. But that's not how aircraft fly! They start, land, and fly in different directions and at different altitudes—and then the separation is not always at the same location. Solid vortex generators then disturb the flow unnecessarily and can sometimes make things worse. It turns out we need to be able to not only control the flow separation but also when and where our vortex generators are turned "on."

Luckily, there is another solution that is helpful in such situations: Instead of covering the aircraft in solid vortex generators, I found scientific papers that showed how we could use tiny jets of air to generate the helpful vortices. The tiny jets make the flow "stumble," creating a small vortex. We decide when and how these air-jet vortex generators operate so that they only influence the flow when and where we need it.

CURIOUS GIRL NOTE—How can we reduce flow separation?

- Use vortex generators (small arrow or ramp-shaped devices) on the surface in front of the flow separation.
- Vortex generators only work if aircraft fly in a straight line, so we need to control when to turn them "on" and "off."
- Use tiny jets of air to create vortices!
- Why does that work? We don't know—maybe the jet diameter, air pressure, or direction?

There still was a mystery to solve. We knew that air-jet vortex generators can very effectively reduce separation in laboratory experiments. But no one knew exactly why.

There are many different parameters that influence the ability of the jets to control the flow—the diameter of the air jets, the pressure at which the air is injected, the direction into which the jets are injected, to name just a few.

We needed to understand their influences.

Also, you probably can imagine that to control separation on an entire wing or space launcher, more than one air-jet vortex generator is needed.

So, in addition to all the parameters influencing each individual vortex generator, we also needed to find out if and how they influenced each other. There was a lot of work to be done before I understood why air-jet vortex generators work so well in controlling separation.

One of the important components of scientific research is to learn what answers to your questions other people have already figured out.

After reading many scientific papers, I had learned that no one had studied how the spacing between the jets influences how well they control separation. Excited to find a research area where I could make a real scientific impact, I came up with a configuration my group and I could study to find out about the effect of spacing.

For tests in a wind tunnel, we always reduce the problem to the simplest form we can imagine and analyze the parameters one at a time. If we were to change more than one parameter at the same time, it is impossible to know what influence each of them has.

To keep things as simple as possible, we did not use the model of an entire aircraft with curved surfaces and wings. Instead, we used a flat plate (basically a flat metal board) and attached a ramp to create a shock wave and a separation region. After manufacturing the model in the mechanical workshop, we installed it in the wind tunnel and ran a first set of experiments to visualize the flow, find out where the flow separates, and where the shock wave forms.

Once we knew where the separation would appear, we could place a row of air-jet vortex generators in front of it and start the actual experiments. To analyze the effect of spacing between the jets, we designed several rows of jets with different spacings that we could place in front of our separation (one row at a time). Between measurements, we only changed the spacing between the jets by installing a different row; everything else was kept exactly the same.

We used several different measurement techniques: flow visualizations to find out where interesting things happen on our model, and then—only in those places—more time-consuming pressure and velocity measurements.

From these experiments, we learned that the spacing of the air-jet vortex generators had a significant influence on their ability to control separation. For large spacings, the separation was reduced. (Yay!) For intermediate spacings, it was reduced even more. (Double yay!) For small spacings, the separation increased! (Wait, what? We actually made it worse?)

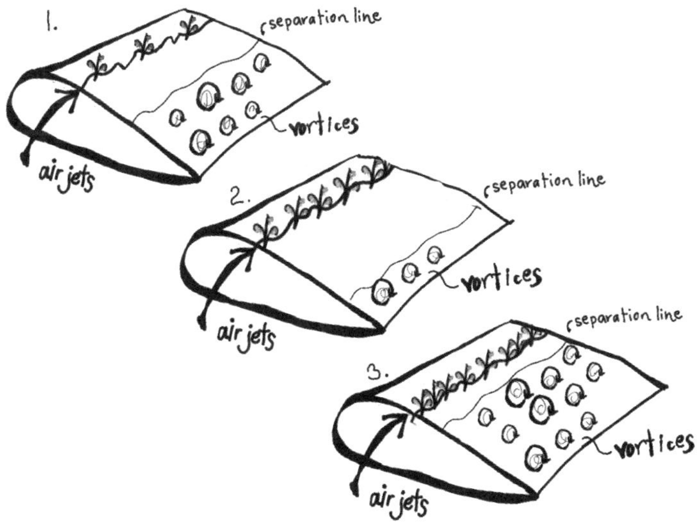

CURIOUS GIRL NOTE—Hypothesis: The spacing between air-jet vortex generators is important.

- Large spaces reduce separation.
- Medium spaces reduce it *even more*.
- Small spaces increase separation? Why? Small spaces interact and disturb each other. Weird . . .

This result made me very curious, and I started to wonder. *What if the vortices generated by the air-jet vortex generators interacted with each other? And by doing so, changed their influence on the flow?*

We could unfortunately not directly see any interactions between the jets from our experiment because the goal of the original experimental design was to give as a full view of the entire flow field—from the air-jet vortex generators through the separated flow region. With this experimental setup, we could not measure the details of the flow around the tiny jets and in the pipes through which the jets are injected, which have a diameter of one millimeter—the same as the point of a sharp pencil. And we could definitely not measure how they change over time. These new questions required a whole new setup.

Experiments in such thin pipes are almost impossible. But to prove our hypothesis about interactions between the jets having a strong influence, we needed a different kind of data that was able to resolve the flow in the thin pipes and the vortices generated by the jets.

So we used a different kind of diagnostic tool: computer simulations. With the computer simulations, we could study how the flow in the pipes and jets develops over time (like under a strong microscope), and I could finally observe the behavior of the vortices in the flow in as much detail as we needed.

And our hypothesis was correct!

When the jets are placed close enough to each other, neighboring vortices start to interact in a positive way, and that allows the jets to influence the flow downstream of themselves and their neighbors—they influence the flow along the entire width of the model instead of only downstream of themselves. This is why multiple jets reduced the separation much more than a single air-jet vortex generator.

But if they are placed too close together, the neighboring jets disturb each other; their interactions hinder the formation of the vortices, making the separation worse, just as we had observed in our wind-tunnel experiment but had not yet been sure why.

By understanding the big influence of jet spacing, we not only identified the influence of one of many parameters, we also found the big difference between air-jet vortex generators and the more common mechanical vortex generators. Only the jet orifices are small enough to be placed so narrowly together that the generated vortices can interact. Mechanical vortex generators (e.g., in the shape of small ramps) need to be bigger to create vortices of the same strength, so no matter how closely they are spaced, the vortices they create are always too far apart to interact.

We were quite happy that our approach and work process had paid off.

Learning from this experience, we always combine different methods to answer our research questions. Combining different experimental methods with numerical simulations provides different bits of information about our flow field, separation, and the jets for flow control. Like in a puzzle, where the right pieces at the right location help create a bigger picture. It is exciting to add these pieces of new knowledge and to contribute to the understanding of our world one little step at a time.

I do enjoy this kind of work very much because I work with a great group of motivated and curious people. In addition to my own research group, I get to meet and work with many other interesting people from all over the world. I can ask—and if all goes well, also answer!—my own questions. And I have so many questions and ideas on fascinating topics that advance the development of new air and space technology—with a mindful use of resources.

from sparks to purpose

The Story of a Girl Who Is Curious about Fire

RITAMBHARA RAJ DUBEY

Every holiday season, my parents, grandparents, and all our families plan a vacation.

This year it was going to be something special because, as part of going to my grandparents, we were going on a trip to the Himalayas. This would be my first hike in the Himalayas, and we were going to hike across the Hampta Pass, which connected a lush green valley with a barren mountain, at an elevation of fourteen thousand feet.

As we drove to my grandparents', we passed a coal power plant during sunset. Contrasting to the oranges and yellows of the sky was the black smoke coming out of the plant's chimney.

I turned to my elder brother. "Why is the smoke black?"

"Probably because the emission control equipment is not installed," he said.

"What does that equipment do? Why would it make the smoke black if it's not there?" I asked, confused. I had a close bond with my elder brother, even though he was much older than me. He always shared with me what he was learning in school and his courses.

"It's a filter," my brother said. "It keeps harmful stuff, like soot and pollutants, out of the air. Without it, all that dirty smoke just goes straight into the sky. These pollutants can be really bad for the environment and the health of the people living nearby."

"But why do they do it, and how about the people living close by?" I asked.

"They do it to save on cost, and it negatively affects people's health," he said. "You can become an engineer or scientist, learn more about it, and then make the people aware of it."

The quote my grandmother recited every morning ran through my mind: "*Sarve bhavantu sukhinah, sarve santu niramayah.*" It means, "May every creature be happy and free from suffering."

"Well, I don't want to be an engineer or scientist, but I just want to help these people," I said.

"You can do both!" said my brother.

That night at my grandparents' house, there was an undercurrent to all of the stories and laughter as I overheard my parents talking in hushed tones. I heard the words, "Maybe we should cancel," and my heart dropped.

"Is everything okay?" I asked.

"No, dear," said my mom. "There is a wildfire in the same area where we are planning our hike. A lot of people have been evacuated, and some have lost their homes. The smoke is making it very unsafe to breathe."

"You mean like the black smoke from the coal power plant?"

She nodded. "It will affect the air quality and our health. It's just not safe to go."

Disappointed, it was hard to accept the change in plans, but I understood that safety should come first. That night, the news of this fire was everywhere, making me curious.

I went back to my elder brother with a thought. "We use fire for cooking, for keeping warm in our houses, and during camping."

I continued, "But it can cause problems like what happened during this wildfire. It's like a fire has these two sides—helpful and not-so-helpful."

We talked a lot that night about how fire could be both good or bad, depending on the circumstances. Though I was disappointed about the canceled trip, my curiosity about fire and smoke stuck with me.

Years later, as I watched a video on using prescribed fires in an environment and ecology class, I started to see the ways that fire could help the environment.

I learned that prescribed fire is an example of using fire as a tool, much like cooking and keeping our homes warm. Prescribed fires are planned, ignited, and monitored by experts in a predetermined area. They help remove dead, dry forest fuels; pests; and diseases from an area and encourage new growth. In fact, I learned that some trees, like the sand pine tree, need the warmth of fire to heat up the pinecones and to let the seeds escape from them to grow into a new tree.

In class one day, our professor asked, "How do we start a fire?"

"Using the fire triangle," I said. "Fuel, oxygen, and a heat source."

CURIOUS GIRL NOTE—How do you start a fire? Use the fire triangle. You need:

- Fuel
- Oxygen
- Heat source

"Absolutely correct. One needs all three of these components to start a fire." The professor paused. "What happens if we take away one of these components?"

"Just like the absence of one wheel from a tricycle, it would prevent the fire from moving or spreading."

"This is the goal of the prescribed fires we are learning about—reducing the fuel in the targeted area to reduce the chances of destructive wildfires later."

I wondered out loud, "Can prescribed fires be harmful like wildfires?"

"Like any other fire, prescribed fires generate smoke that can travel thousands of miles from the location of the fire and impact air quality and people's health if not managed properly," the professor explained. "Smoke is made of tiny particles and gases from things like wood or grass that didn't burn completely. These particles are so small—about twenty-four times thinner than a human hair—that they can travel through the air and even get into our bloodstream if we breathe them in. Besides this, the smoke can cause poor visibility in nearby areas. This is why we need to have an understanding of the smoke's path before igniting these fires. By understanding how smoke spreads, experts can plan prescribed fires at the right time and place to avoid affecting nearby towns or air quality."

CURIOUS GIRL NOTE—Impacts of fires on us and our environment:

- Destruction of property
- Dangerous spreading (up to thousands of miles)
- Air quality
- Health: smoke particles can get into our lungs and bloodstream (particles can be twenty-four times thinner than a human hair!)
- Reduced visibility due to smoke

As part of that same class, I met a fire safety scientist, Dr. Kalpana. As I listened to Dr. Kalpana's experience of studying fires and handling them, my heart raced. She participated in controlled burn practices, observed fire behavior, and volunteered for the local fire department in her free time. Listening to her, I realized that Dr. Kalpana's passion for her work was contagious. I found myself drawn to the idea of being in a profession like hers where I could learn cool things and also help the community and ecosystem.

Now donned in firesafe clothing, preparing to record measurements amidst a prescribed burn as a researcher feels surreal. The crackling flames and the smoky air serve a vivid reminder of how far I've come and how much more there is still to learn. I am still grappling with the complexities, trying to understand how something as simple as moisture in fuel can drastically change the nature of smoke—its composition, dispersion, and impact on nearby regions.

I also feel a sense of responsibility, realizing that each observation and data point I collect brings me closer to unlocking these mysteries. I've learned that the composition and movement of smoke are not just about science—they're about survival. This knowledge goes beyond academic curiosity; it's about protecting lives and ecosystems.

As I quietly watch the fire, I can't help but think how my childhood curiosity ignited a path of purpose. Looking back, my fascination with fire as a child set me on a path of purpose. That thick black smoke and wildfire haze are more than pollutants—they're challenges I now work to understand and manage.

My research is dedicated to these very issues—to mitigate the harmful effects of smoke and lessen the environmental impact of fires. I've transformed from a curious child questioning my elder brother about black smoke to a scientist tackling these challenges. Fire is powerful, and understanding it can help us protect both people and nature. All it takes is a little curiosity—and the courage to ask questions.

wonders flowing in our veins

The Story of a Girl Who Is Curious about Blood

CHRISTINA RICE

I've always been fascinated with how the human body works. Have you ever fallen and scraped your skin? Did you see how the blood that came out eventually dried up and sealed your wound? Blood constantly flows inside our bodies to seal wounds, move around the oxygen we breathe, and even help heal sickness.

Have you thought about how blood moves inside us? Because it's red, maybe you think of ketchup or Kool-Aid. But our blood is more like a soup. Like a soup carrying peas and carrots, our blood is a thin liquid carrying different cells. When our bodies are healthy, each cell in our blood can reach its destination. But when our bodies are sick, these cells can change and have difficulty getting to where they need to be.

CURIOUS GIRL NOTE—How does blood move through us?

- Blood is *not* homogeneous like ketchup; it's more like a soup, a thin liquid with different cell components (like peas and carrots).
- Blood flows through vessels—different-sized tubes—from the heart outward.

I did not understand all that happened in our blood as a young girl, but I did wonder how the world worked around me. I asked lots of questions and learned I liked solving problems. At first, I was interested in the math and science courses during grade school. Then in high school, I took a technical drawing class because I did not want to paint. The class was designed for engineers, who I learned use math and science to solve specific problems.

I kept taking elective classes focused on engineering. I joined a club team that built machines and robots. I worked with classmates to draw plans, manufacture parts, and assemble working devices. These projects helped me decide that I wanted to attend enough school to get a job where I got to ask novel questions every day. I wanted to spend my career seeking answers that explained how the world worked and how to leverage knowledge to benefit our world.

After starting university to study engineering, I had the opportunity to intern at a nonprofit overseas. The nonprofit—located near the city of Gulu, Uganda, in Eastern Africa—contributed to health efforts such as water access and safe cooking fires. In an effort to provide clean water to a nearby village, we were digging a well. The heat beat down on us while we dug out the soil to reach fresh water. I felt tired and lightheaded. I had muscle aches throughout my body. I had to sit down and stop digging to rest. I told my friend as I drank water, "I am so tired. I don't know why because I normally can complete physical tasks."

He responded, "You should get tested for malaria. Maybe you got bit by an infected mosquito recently."

My tiredness with aches could be a sign I had the malaria parasite in my blood. This disease, found mainly in tropical regions, can make people sick, especially children. Regardless of age, it is crucial to seek medical assistance to help treat the symptoms.

At a local clinic, I watched a doctor take my blood—the red liquid slowly filling the little vials. The lab technician looked at the smear of blood under a microscope, trying to determine if I had malaria.

My curiosity drove me to ask, "How can you tell if I have malaria by looking at the cells under a microscope?"

"The red blood cell shape tells me that malaria is not present, which means you are feeling sick from something else and will likely feel better soon."

The doctor thought my working in extreme temperatures for so long without taking enough precautions most likely led to heatstroke. He told me to drink water and rest for the remainder of the day.

While resting, I wondered how that diagnostic came to be. How was it that the doctor could look at the shape of my cells to see if I was sick? What do healthy red blood cells look like, and how do they change if infected with malaria?

That experience sparked a deeper curiosity about the connection between the symptoms I felt and the science of blood, leading me to study malaria in graduate school. Infected by curiosity, I wanted to know how infected red blood cells change as the malaria infection progresses and how red blood cells travel through the blood.

Blood in the body flows through vessels, or pathways made up of different-sized tubes, from the heart outward. Think back to the soup analogy and imagine pushing soup through tubes. Soft vegetables, like peas, may go through the tubes more easily than harder vegetables, like undercooked carrot pieces. A healthy cell is more like a pea that can be squished, while an unhealthy cell is more like an undercooked carrot that is difficult to squish.

In large vessels, this means the healthy and unhealthy cells pack together differently based on how easily they can change their shape. As the vessel narrows, unhealthy cells that are hard and swollen may not be able to enter the pathway at all.

Not being able to move through the vessel makes it more difficult for the oxygen carried by the red blood cells to reach the whole body. This is one of several reasons that people with malaria will feel sick. Other reasons include how the unhealthy red blood cells push other cells traveling with them in the blood away from where they can best perform their function, and how these infected red blood cells can stick to the vessel walls to avoid the body's natural defenses.

Why do the malaria infected red blood cells change shape? When the malaria parasites enter our bodies, the parasite searches for a red blood cell and hijacks the internal components of the red blood cell to make copies of itself. This damages the cell, making it swell and resist bending.

While we know why they swell and resist bending, I needed to know *why* these swollen cells behaved so differently than unswollen ones. What was the impact of that change?

I started by closely examining how malaria swells red blood cells. A healthy red blood cell has a unique shape that we refer to as a biconcave disc, which resembles a frisbee with a thick lip all the way around and a dip in the center from both the top and the bottom.

A malaria-infected red blood cell will swell. The frisbee lip thickens, and the center dip gradually disappears. A greater change in shape indicates that the parasite has made more copies of itself as part of the infection. Eventually, the infected red blood cell will pop, and the parasite copies will move on to infect other healthy red blood cells.

CURIOUS GIRL NOTE—What does malaria do to blood?

- Diseases like malaria change these properties of blood, like red blood cell (RBC) shape and flexibility.
- RBCs change shape because parasites hijack parts of the cell (damage)!
- These parasites cause RBCs to be swollen and less bendable, eventually bursting to release more parasite copies.
- Unhealthy cells might not be able to enter smaller vessels → hard to get enough oxygen!

Back in Uganda when my lab technician told me that they could tell by the shape, they were looking for these swollen discs with parasites inside. Fortunately for me that day, my blood sample contained discs with center dips and no parasites.

It turns out that the swollen shape of infected red blood cells does more than give doctors a way to determine if the parasite is present. It also helps the parasite avoid the body's natural defense system. Our bodies have many ways to defend themselves from invader parasites. One defense is the spleen.

The spleen is not an organ we hear about as much as our heart and lungs, but it is pretty important in keeping us healthy. The spleen's job is to filter red blood cells that cannot bend from our blood. Only healthy cells that can bend and pass through a series of narrow slits inside the spleen get to continue through our body.

Many infected red blood cells won't even reach the spleen because they stick to the blood vessel wall. Before I tell you why this happens, I must tell you about healthy cells first.

Healthy red blood cells often avoid colliding with blood vessel walls. Researchers observe that blood flowing through midsized vessels forms a core of densely packed red blood cells surrounded by a layer near the wall with few red blood cells. Blood platelets and white blood cells prefer to flow in this open layer near the wall to best complete their jobs in the body. This can be pictured as a street with bike lanes. The red blood cells are the cars using the center of the street, and the other cells are using the bike lanes. Infected red blood cells resemble cars entering the bike lanes where they are not wanted.

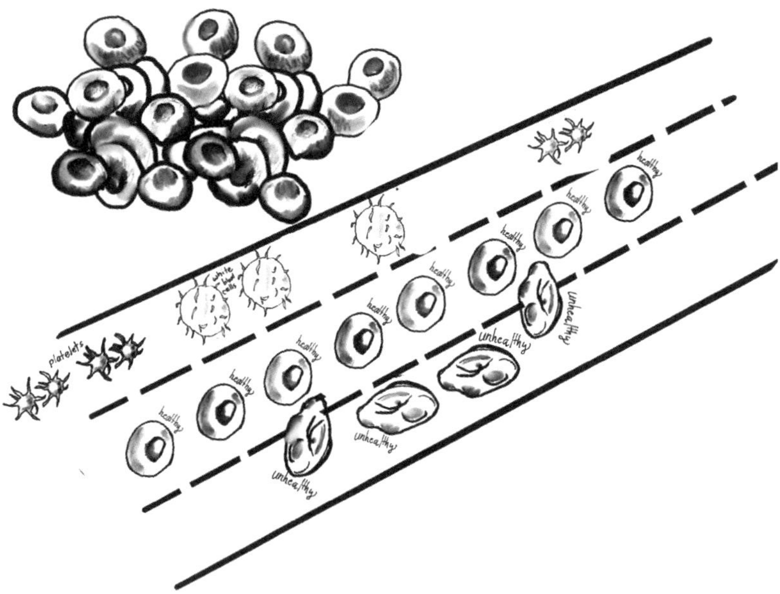

But why would swelling and hardening make the infected red blood cells move toward the vessel walls? First, imagine a kite flying. Some kites are better at flying than others. For example, a good kite will lift upward toward the clouds, while a bad one will nosedive toward the ground. The change in shape can force the infected red blood cell to lift toward the wall.

Second, imagine trying to pack together soft discs made of stuffing versus hard discs made of rubber. It is easier to tightly pack the soft discs together. Hardening the discs can make it harder to pack them densely together with other cells. Once it is difficult to pack densely, it is easier for the cell to travel outside the core nearer to the wall.

This series of images gives us a general understanding of how blood flow changes when people are infected by diseases like malaria. The healthy blood is made of red blood cells taking the shape of dimpled frisbees and traveling largely in the center of blood vessels.

Malaria-infected blood has a mix of dimpled and swollen frisbees with infected red blood cells traveling closer to the vessel walls.

Understanding this led to the next step of my project—exploring how I can create models that can be used to generate ideas about how to reduce the impact of malaria-infected red blood cells.

To run my experiments, I use computer simulations rather than taking blood from someone. When I first began computations, I checked that my program was accurate compared to experiments ran with blood samples from volunteers. The computer program allows me to explore a range of possibilities without the need for needles!

The computer program I wrote imitates the movement of blood cells through plasma—the liquid of our soup. My program not only has healthy and unhealthy red blood cells but it also includes white blood cells (cells that fight disease) and platelets (cell pieces that help us stop bleeding by forming clots). It cannot include every aspect of our real-life blood, so we strategically select the most essential aspects to replicate on our computers.

Choosing what information to include and exclude comes with a lot of trade-offs. We know we need to include information about how the cell changes shape and stiffens, but what else? And how will we represent shape change and stiffness in a computer program?

We capture shape change by uploading 3D pictures drawn on the computer into the code. These 3D pictures can represent both cell shape and the shape of the tube the cells travel through.

Capturing how squishy the cell is can be more challenging because the cell shape needs to change as cells push against each other. It is not a single frozen picture but a series of steps unique to each collision. We have addressed this by covering the surface of the cells with many connected springs.

Springs are commonly used in science models because they want to return to their resting positions. If a spring is shortened while bending, it will push against the force, shortening it as it tries to return to its resting position. The opposite is also true; if a spring is elongated, it will pull against the force, elongating as it tries to return to its resting position. This means that the cell shape will be able to bend when colliding with another cell or wall, but once the collision is complete, the cell will return to its original shape.

CURIOUS GIRL NOTE—How can we tell the difference between healthy and unhealthy RBCs?

- Healthy: Cells don't collide with blood vessel walls.
- Unhealthy: Too swollen and hard to enter tiny vessels, so they collide with the walls *a lot*.
- We can model healthy versus unhealthy cells using computer simulations, including cell shape, vessel shape, and cell squishiness (modeling cells by covering their surface with springs).

How strongly the spring fights to return to its original position can be tuned by the engineer. If we program a strong spring, it will resist changing shape, and a weak spring will easily change shape. This ensures that the healthy cells with weak springs are soft and squishy, while the infected cells with strong springs are hard and rigid. Combining the fluid holding the cells, the differently shaped cells, and the spring-covered surface of the cells completes the first version of our computer model.

We further develop the computer model by working with doctors who treat malaria patients. We hope to use the model to find a trend or pattern about how infected red blood cells position themselves in our body. This knowledge will inform medical doctors as they work on better treatment methods and explore new drugs to target cells stuck to the vessel walls.

It may sound like science fiction, but biomedical engineers are already designing drug delivery "vehicles" that can carry a drug to specific parts of the blood and body. These "vehicles" are tiny particles that travel through the blood, often near the wall using the bike lanes of the vessel, with medicine slowly flowing out from within the particle.

When this project finally comes together, a doctor could select the appropriate vehicle to deliver the most helpful dosage to the most impacted region using the information found in our model.

Just imagine one day in the future when a young girl living in a tropical climate comes down with fatigue and muscle aches. Her parents take her to the local clinic, and the doctor draws her blood to test for malaria.

On this day, when the doctor looks at her red blood cells under a microscope and sees swollen discs, they can offer our sick young girl a shot that has small particles that travel through the blood strategically to deliver medicine to the infected red blood cells. These particles will be picked based on how long she has had the parasite in her system and where in her body the unhealthy cells are traveling. The young girl leaves the office with a skip in her steps, confident she will feel completely healthy soon.

This day may still be in the future, but every step we take brings it closer.

As a girl, I did not know how curiosity would shape my life. Asking questions has led me to learn and experience incredibly beautiful things during my life. It also introduced me to many great people who taught me how to explore the world systematically and communicate all the thoughts in my head. Asking questions fueled my path from a girl willing to observe a scraped knee instead of looking away to an engineer who travels the world to work with doctors to make better simulations of blood.

I did not know exactly where it would lead, and I still am surprised every day by what comes next, but I keep asking and exploring the world around me. Curing a disease like malaria takes more than one person and plenty of time. It involves many types of questions—from scientific to political to cultural. But as we learn, we generate hope—hope that we can understand better, to heal better.

Dynamic stall phenomenon happens when an object moves quickly or dynamically in the air. The phenomenon happens due to the wind-object connection. When an object is in the air, the air surrounds it, and the air molecules cling to the object's surface. When the object moves quickly, the air molecules find it challenging to stay attached and follow the object at the same speed. The molecules detach from the object's surface, swirl, and get chaotic. The object will suffer from continuous chaos and unsteadiness. This will cause the object to shake, and if the intensity of shaking is high, the object might even break.

mysterious forces in the wind

The Story of a Girl Who Solves a Mystery about a Wind Turbine

SAHAR REZAPOUR

"I want to become a detective" was my immediate answer to "What will you be when you grow up?"—an unusual response that often raised people's eyebrows in surprise and curiosity. As a kid, I couldn't get enough of detective stories in books and cartoons. Hours would slip away while I was doing puzzles, spotting differences between two seemingly identical images, and figuring out tricky riddles. My dream of becoming a detective evolved over time. The love of solving enigmas and mysteries combined with my passion for science led me to become a detective in science—a researcher—to figure out scientific mysteries and explore the unknowns of nature.

What does my job look like?

It all starts when a mysterious case comes our way, leaving us scratching our heads and wondering how and why it happened. Our first stop is to visit the archive—a treasure of books, articles, and reports from other detectives that hold the secrets of similar events from the past. Some potential suspects might come out from the archive. We will dig deeper and try to reconstruct the crime scene—the event—in a laboratory or with a computer to gain more knowledge about the event. We put all the information together to find the answer to the mystery and detect the cause—the culprit. This information will help other scientists find solutions to prevent the event from occurring in the future.

Among all the mysteries in the world, I am particularly curious about the invisible force of nature—the wind—and how it interacts with objects. Wind plays a big role in many of my childhood memories, whether it was chasing butterflies, watching birds fly, blowing bubbles and dandelion seeds in the air, or playing with pinwheels and kites. This passion fueled my desire to become a detective in the science of understanding how the wind and objects work together.

One day, while working in the Wind Quest scientific laboratory, I heard the sound of something being dragged along the ground. It was a different sound, unlike the usual noise of the cleaning machine that tidied up the corridor once a week or the random drilling noises from the never-ending random constructions. Grumbling about the never-ending noises in the building, I heard a knock on the door.

"Hi, I'm Hannah, and I believe I've come across an extraordinary scientific mystery relevant to this laboratory." She was probably the youngest visitor I'd ever had in this office.

She was carrying an object almost twice her height. Definitely not a toy, and definitely not resembling any other object I'd seen before.

"This is a wind turbine that broke unexpectedly, and I need your help to figure out why it broke," she explained.

As someone working on wind-related objects and being proud of having read more than ten thousand papers about wind turbines, I have to admit, I had never come across a wind turbine shaped like it. Wind turbines were giant metallic pinwheels, but this one had a body similar to the trunk of a tree with only three branches, each branch attached to a large rectangular plate. It could be more of a creatively designed rotating coat tree, but definitely not a wind turbine.

I was preparing sentences in my mind to appreciate her effort and enthusiasm, thank her for her visit, and regret to inform her that I might not be able to help.

Before I could say anything, she continued, "You probably don't think this is a wind turbine, but I do. In fact, if you see it as just a coat tree, you have every right to. That's actually how this wind turbine was designed."

Her directness left me speechless. Setting aside my initial impulse to dismiss her, I grabbed my notebook to jot down the incident details. I decided I would figure out later whether to take on her case.

Hannah came from a family with a long-standing tradition of building wind turbines; the great-grandparents were the founders of the business. Inspired by pinwheels, they created turbines to generate electricity. Their innovative approach became a huge success, helping to reduce reliance on burning fossil fuels like oil and gas. Fewer toxic fumes were polluting the air, and the environment was less threatened.

As generations passed, the great-grandparents' legacy continued to thrive. Their descendants kept improving the business in pursuit of making the most from wind energy. A significant limitation was preventing these giant pinwheels from unlocking ultimate energy production. Like pinwheels, those turbines always needed to face the wind directly to spin and generate electricity. They would struggle to turn well if the wind blew from the back or side, and improper rotation meant insufficient electricity. The family, although successful in the business, always wondered if they could overcome this limitation.

Hannah had answered this question during a childhood game. Her favorite hobby involved placing a spinning coat tree in front of a ventilator fan, hanging from the coat tree, and spinning as the fan blew on her. She had noticed that no matter where the fan was positioned, she could always spin. This led her to wonder: Why not design a turbine that worked the same way, with rectangular plates attached? If her new design proved successful, wind turbines could become even more efficient at generating electricity. Ultimately, Hannah dreamed of making a significant impact in saving the planet.

The construction of the new turbine began in the factory's workshop, and she dubbed the turbine H-shape. In the factory, every newly built turbine underwent a rigorous testing process to ensure it was ready for use and capable of generating electricity efficiently; H-shape was no exception. The team was eager to see how well the new turbine compared to the company's long-standing models. So they decided to turn the simple assessment test into a tricky competition. They gathered the finest turbines the company had ever produced on the test field. The objective? To determine which turbine could generate the most electricity. They measured each turbine's performance by the number of light bulbs it could illuminate. H-shape was a pioneer in this competition and broke the company's electricity generation record, lighting up an impressive 9,488 light bulbs all on its own!

Everyone burst into cheers, with eyes filled with anticipation of how the future of the business would be with this new design, how much profit it would bring, and more importantly, that they could now fight better to convince others to stop burning oil and gas. The fact that this turbine was generating such a large amount of electricity in almost complete silence was a major bonus. Unlike the noisy pinwheel-type turbines that scared away the birds, this turbine offered a quieter alternative.

All of a sudden, however, H-shape began to wobble, emitting a noise like a cry for help. Before anyone could react, one of the arms snapped off and crashed to the ground. All the other turbines stood firm, showing no signs of wobbling or vibration. The crowd fell into stunned silence, unsure of how to react.

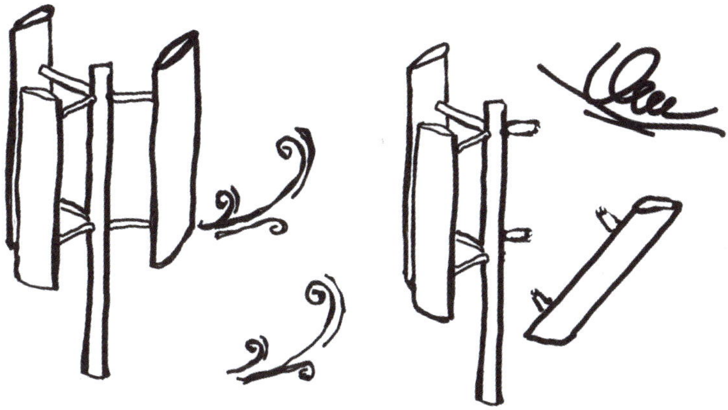

I stopped taking notes. She took a deep breath, shifting her gaze toward H-shape in the corner of the office, and continued in a softer voice, "I felt hopeless. I gathered the broken pieces and returned home without seeing anyone from my family. My dream of saving the planet had shattered, and with it, the hope of my family's business development. I hadn't ever fallen when I was playing with the coat tree. But then why did it happen to our H-shape? I couldn't let it end like this."

No one had wanted to give it another try. Even the workshop had refused to help her repair it. So she had begun searching for alternative places to fix it, and coincidentally, she had come across our lab's webpage.

"At the Wind Quest lab, we explore the relationship between wind and objects."

When I heard her story, I couldn't help but get intrigued. I had no idea what might have happened to her turbine, but it was a unique opportunity for me to practice some detective skills. I took her to my colleagues to start the first step: reviewing the archives. We began gathering clues and information through countless books, articles, and online sources, desperately seeking any trace of objects resembling the enigmatic H-shape. Hiding my frustration was getting more difficult, as every lead seemed to lead to a dead end.

Instead of continuing to search for similar turbines, I decided to focus on unusual wind-related incidents. In the depths of forgotten archives, I stumbled upon an old document—a research paper about a mysterious phenomenon that could shake and even break the main blades of helicopters.

This phenomenon was called dynamic stall. *Stall*, a fancy word for stop, and *dynamic stall*, the phenomenon that stops moving objects.

> *Dynamic stall* phenomenon happens when an object moves quickly or dynamically in the air. The phenomenon happens due to the wind-object connection. When an object is in the air, the air surrounds it, and the air molecules cling to the object's surface. When the object moves quickly, the air molecules find it challenging to stay attached and follow the object at the same speed. The molecules detach from the object's surface, swirl, and get chaotic. The object will suffer from continuous chaos and unsteadiness. This will cause the object to shake, and if the intensity of shaking is high, the object might even break.

CURIOUS GIRL NOTE—What is dynamic stall?

- Has to do with the wind-object connection.
- Object moves so quickly in the air that air molecules can't follow at the same speed.
- Molecules detach from the surface, swirl, and become chaotic. This is dynamic stall.
- Object suffers unsteadiness due to the chaotic flow.
- Intensity of shaking may be high enough to break the object.

My heart was filled with renewed determination and hope, as the problem was now solved for the helicopters.

As I read more about this topic, I found even more interesting facts. I was fascinated to learn that birds and insects like butterflies use a similar mechanism to fly by moving their wings fast enough. This mechanism makes these animals more powerful and lets them go higher in the sky. H-shape had also been remarkably powerful, breaking the electricity production , but it couldn't sustain the power. The birds and insects knew how to move their wings quickly while keeping the power.

They don't let the air molecules separate suddenly. They skillfully manage their interaction with the air.

"Are you sure this is what happened to H-shape?" Hannah asked, her skepticism clear.

I hesitated, unsure myself. After all, H-shape bore no resemblance to birds or helicopters. "The only way to truly understand is through experimentation," I replied.

We transported H-shape to our research lab to investigate whether dynamic stall had caused its failure. With great care, we set up a special experiment to recreate the conditions that led to H-shape's vibrations.

At the heart of our lab is the wind tunnel—a large, cubic channel with powerful fans to create artificial wind at various speeds. This wind tunnel allows us to simulate everything from gentle breezes to brutal storm winds.

The key was to observe the air particles and how they move around H-shape's plates. One of our main challenges was to make the invisible air molecules visible. We visualized the air particles with the particle image velocimetry technique. To use this technique, we released tiny particles, like dust or glitter, into the air. The particles should be tiny and light to float in the air. We illuminated these particles with a green light. The green light caused the particles to glow brightly, allowing us to easily track their movement and study the behavior of the air particles. We recorded long videos and took numerous photographs to capture how the particles behaved as they interacted with the turbine.

CURIOUS GIRL NOTE—How can we "see" the wind with particle image velocimetry?

- Release tiny particles into the air (particles must float).
- Shine bright light (a laser) to light up the particles.
- Track the particles using the glow from the laser light illuminating them.
- Take pictures and videos to capture how the particles in the air move.

The truth immediately disclosed itself as we scrutinized the captured images and videos. Dynamic stall was undeniably the cause behind H-shape's failure. The images vividly displayed the dramatic separation of air particles from the turbine's plates' surface during the vibrations.

Hannah's voice broke the silence with a question: "But why did this phenomenon occur? What causes it?"

I began repeating one of my childhood hobbies: staring at images, carefully examining them, and spotting their differences. The images finally uncovered two primary suspects responsible for H-shape's crisis: changes in wind intensity, and changes in wind direction.

The fluctuating wind intensity proved to be a major factor. The wind pushed and pulled the turbine in varying manners. The constant change between pushing and pulling subjected the air particles to strain, causing them to lose their grip on the surface.

The other reason was the changes in the wind's direction. When the wind blew in the same direction as H-shape's plates, the tiny air particles maintained their connection and flowed in order over the plates. When the wind hit the plate in a highly tilted direction, the air particles couldn't follow the surface of the plate, and the particles would separate. Immediately following the separation, H-shape would tremble uncontrollably.

Hannah's designed turbine possessed the extraordinary ability to harness wind from all directions, but this superpower came with a unique challenge. A drastic change in wind direction was unhealthy.

"I understand what happened. But how can we use this information to make H-shape more powerful and not let it fail?" I wondered aloud.

Hannah, the little spinmeister, eagerly said, "I think I know what it feels like! It's like . . . it's like holding an umbrella on a windy, rainy day. The turbine should *dance* with the wind! Both wind intensity and direction can cause air molecules to separate. We cannot change the natural flow of the wind. The solution is to constantly adjust the plate's direction. But, how can we make the turbine blades dance wisely?"

The answer was already given in Hannah's question: We needed to make the turbine smarter! The turbine had to be aware of the surroundings. It had to sense what was happening. So the plan was to add sensors on the plate. These sensors would sense and detect vibrations and immediately transmit signals to small computers. These computers would serve as the turbine's brain, automatically adjusting the blades to dance harmoniously with the changing wind patterns.

So far, this approach has helped H-shape to stay balanced and effectively harness the power of the winds. Hannah's family began using this technique for all the future turbines, making wind power accessible to more people worldwide.

I finished H-shape's case report and archived it, hoping it might inspire the discovery path of a future mystery by another detective.

But submitting the report doesn't mean the case is closed. For detectives in science, no case ever truly closes. Each resolution opens the door to new mysteries waiting to be solved. As a curious detective, I'm still eager to uncover more about dynamic stall and the breaking of the H-shape's arm. We've identified a team of suspects, each contributing to the puzzle, but the true satisfaction will come from identifying the leader behind it all. Is it the sudden changes in wind direction that plays a key role? Or is it the variations in wind intensity? Or could it be the turbine's shape?

So many questions, and so many uncertainties, but one thing I'm certain of: My investigation adventure is far from over!

breaking waves and derailed dreams

The Story of a Girl Whose Life Was Changed by the Ocean

KELLI HENDRICKSON

"Finished feeding the fish?" a voice over my shoulder called in a light, singsong voice.

My hands shook as I clung to the rail, and I quietly wondered if I had anything left in my stomach to "feed the fish."

"Maybe." I swallowed.

"You should be getting prepped for Dive 2." The divemaster's joking voice had a note of kindness in it.

Ugh. "Only a three- to four-foot swell," the divemaster had said. In the grand scheme of ocean swells, a three- to four-foot swell was like ripples on a pond. Small. Inconsequential. That's how I felt right now.

But also, I felt nauseous.

This wasn't my first dive in open water. I had dived off the coast of Cancun and the Bahamas, so why would diving off the Caymans be any different?

As part of changing my degree from aerospace engineering to ocean engineering, I had gone all in on studying and exploring the ocean. I got my diving certification and entered graduate school planning on incorporating it all into a PhD that studied how waves moved throughout the ocean.

"You should feel better at depth," he said. I watched the divemaster's wet and bare feet turn as he walked away to take care of others in the group. It was a modest-sized boat with an inner cabin, and I could hear him check on each group member in turn.

I gripped the rail tighter and looked out at the ocean before me, unforgiving and indifferent. The nausea churned like the waves beneath the boat. My thoughts drifted back to an hour earlier, when I was eagerly getting ready for Dive 1.

Our boat had slowed to get into position, slowly rocking with each wave that came through. As I sat on the bench, carefully checking my scuba diving equipment so that I could breathe underwater, my meager breakfast of eggs and toast started to make themselves known in the pit of my stomach.

I looked around. Everyone in my dive group seemed to be going about their business and going through the checklist: air tank, dive weights, eye mask, breathing regulator.

My checklist was a little different:

Tank.
Don't throw up.
Weights.
Mask.

Don't throw up.

Regulator.

Don't throw up.

The divemaster noticed me looking a little ill and placed a gentle hand on my shoulder. "It'll get better once you're in the water."

My ocean engineering education told me this was true. That's how waves worked.

The wave velocity decays as you go down in depth. The deeper you go, the less of the surface motion you feel.

I swallowed my pride and my breakfast. *Just make it into the water, and it'll be fine.*

But it wasn't fine.

It didn't matter that I didn't feel the waves move the boat underneath me. My stomach thought I was moving, and it was doing its very best to convince me that vacating my breakfast was the only option available to us.

One of the first lessons they teach you in dive class is how to manage seasickness and dive because it is quite common for seasickness to show up on dive trips.

However, what they teach you isn't, "*Don't* vomit." The lesson is, "*Don't* take out your regulator if you're going to vomit."

If you've ever had the flu so bad you've thrown up, you'll understand. Once you expel the contents of your stomach, the only thing your body will let you do is take a big gulp of air. It's a biological reflex.

Here is what you really learn in that important diving lesson: If you're thirty feet below the ocean surface, and your regulator is in your hand instead of your mouth, your biological reflex to take a big gulp of air will fill your lungs with seawater, and you will drown.

And if you have to throw up, you have to throw up through your breathing regulator.

Remembering this lesson, my stomach and my brain were having this back and forth. My stomach wanted to vacate my eggs and toast. And my brain—well, basically I—did not want to throw up in my regulator. *Desperately*.

My brain won that battle, and I did not throw up at depth. But my stomach won the war. As we got back on the boat and navigated the swell for Dive 2, I spent my time hurling over the rail until the divemaster found me clinging to the rail, "Feeding the fish," as he said.

There was no second dive for me that day. I spent my time waiting for my dive group to return, feeling my plan of being a "scuba-diving, ocean-engineering researcher" disappear a little bit more with every wave that rocked our dive boat.

Back in my office at school, I lamented about the failure of this trip to my office mate. "I have the worst sea legs in human history!"

He quietly typed on his computer and gave me the requisite uh-huhs as I went into detail about how this failure had obliterated my career plans.

Finally, he peered around his monitor at me. "Hyperbole much?"

I threw my hands up, exasperated. "The part that I love most about the ocean relegated me to hanging over the side of the boat, pleading with it to stop. I should have stayed an aerospace engineer. The moon doesn't have waves."

He rested his hand on his chin, studying me. As my office mate for two years, he knew all about my origin story. He knew that, as a college student studying aerospace engineering, I had fallen in love with the ocean on my very first trip to a beach and immediately decided to get a PhD in ocean engineering.

"It doesn't have air or water, either," he said nonchalantly.

I laughed. "No."

Returning his eyes to his monitor, he said, "That's what research is."

I sighed. My office mate was a man of few words, but he had a point. He'd been at this PhD thing longer than I had and had a little more perspective.

"That's what research is" was (one of) our advisor's sayings that was offered up when we ran into challenges with our work. It is shorthand for, "Research is about trying things, failing, and then trying something different." By using our advisor's shorthand, my office mate was telling me to stop complaining about the challenges and find a different way.

He was right. If I was going to be a ~~scuba-diving~~, ocean-engineering researcher, I was going to have to find a way to explore the ocean with all of its dynamic movement with my feet firmly on dry (unmoving) ground.

Before I changed my career direction from aerospace to ocean engineering, the one thing that I really enjoyed was taking the governing equations of fluid dynamics and turning them into mathematical models that I then turned into computer code. I was the only teammate on my senior design team who not only volunteered to but wanted to write a code that predicted the pattern of the compressible shock waves within our engine design. While I worked at NASA, I wrote a code that modeled the water cycle of humans living in enclosed living quarters to try and predict if future Martian astronaut habitats could be sustainable.

For my graduate research, I finally decided I was going to use CFD (computational fluid dynamics) to simulate breaking waves and use the data from those simulations to learn how breaking waves influence the interaction between the air and the ocean—the air-sea boundary layer. CFD was going to be the new way I could explore the ocean with my feet (or basically, my seat) firmly on dry, unmoving ground.

Movies like *Avatar: The Way of Water* and *Moana* make it seem easy to represent breaking waves on a computer.

As I started my newly chosen topic of using CFD for breaking waves, I very quickly learned how hard it was to do it in a way that was physically accurate for even a single breaking wave. Let me explain.

While we think of breaking waves happening near the coast, waves can break anywhere in the ocean. But, for the sake of visualizing, let's think about a breaking wave that you can create in your bathtub. Most bathtubs are about 60 inches long, 30 inches wide, and about 14 to 16 inches high (thanks, Google). Converting this to the metric system with rounding, a generic bathtub is 1.5 meters long, 0.75 meters wide, and 0.4 meters high. The largest wavelength, the distance between two wave crests, that we can generate in our bathtub is 1.5 meters.

Once a wave breaks, there are physics that become important for our bathtub wave! Bubbles get entrained into the water, drops get ejected into the air, and turbulence gets generated. The entrained bubbles can be as small as a micrometer. The droplets that get ejected can be even smaller at half a micrometer.

The computer simulation of our bathtub wave needs to be able to contain 0.45 cubic meters of water volume and be able to see bubbles and drops that are six orders of magnitude smaller in size. This is like taking a selfie of you and a friend on the beach where your camera would need to contain the two of you and capture (in that same photo) the individual grains of sand on the shore, each being six orders of magnitude (or a million times!) smaller than the size of your faces in the photo.

CURIOUS GIRL NOTE—Characteristic length scales for breaking waves in the bathtub:

- Bathtub: 1.5 m long x 0.75 m wide x 0.4 m deep.
- Farthest distance between wave crests in a bathtub is the wavelength, 1.5 m.
- Air bubbles created by the breaking waves entrained into the water are as small as 1 micrometer!
- Droplets of bathwater ejected into the air.
- *Turbulence generated*
- Wavelength is big, bubbles are small!

"We have a length-scale problem." I slunk into my advisor's office and sat in the chair reserved for students during research meetings.

"Okay. Let's talk it out." My advisor greeted me with another one of his well-used sayings.

Together, we thought through all the conceptual arguments on paper and tried to come up with a plan to make the problem tractable. And by tractable, I mean solvable on a supercomputer using days of computer time.

We ended up having to compromise—our smallest bubbles and drops would have to be centimeters, not micrometers. But we agreed that for the bubbles and drops we did include in the simulation, we would get the physics right. Even though I felt like we were missing a piece of our puzzle, I agreed to leave the smaller bubbles as something to return to when I could figure out a way to do those better.

I chipped away at getting my code to work for about a year.

Let me rephrase.

My code worked, and I could make pretty simulations that looked right fairly early in that year. But it kept failing when it came to comparing the simulations results to what I saw in experiments and on my trips to the beach.

In my simulations, the bubbles that were entrained in my breaking waves would come to the surface and *pop!* instantly. Every trip to the wave tank and the beach showed me that bubbles could linger on the surface. After checking and rechecking my code and equations, the only conclusion I came to was that I must have a physics problem.

Determined to get the physics right, I dove into reading up on surface bubbles and found my biggest physics problem—the effects of surface tension and surfactants.

If you've ever washed dishes or taken a bubble bath, you are already familiar with this type of physics.

Until I started reading up on surface bubbles, though, I had no idea how big a deal this was.

When you first fill a sink or tub with water, whatever bubbles that get entrained into the water rise to the surface and burst fairly quickly. High-speed videos of bubbles coming to the surface show us that a thin film of water forms around the bubble as it sits on the surface. Because of the attractive force between water molecules, the water molecules in the main water bulk pull this film back like an elastic band.

For normal water, this force is strong, and the bursting happens in milliseconds. The bubble rises to the surface and *pop*!

Once we add soap, which is a surfactant, the soap molecules crowd their way in between the water molecules and reduce the attractive force between them. For your standard dish soap, this reduction is by about a third. The elastic band that pulls the thin film around the bubble back into the main water bulk is now three times stretchier.

Surfactants (soap) make the bubbles much more stable. The bubble rises to the surface, hangs out for a bit and . . . *pop*?

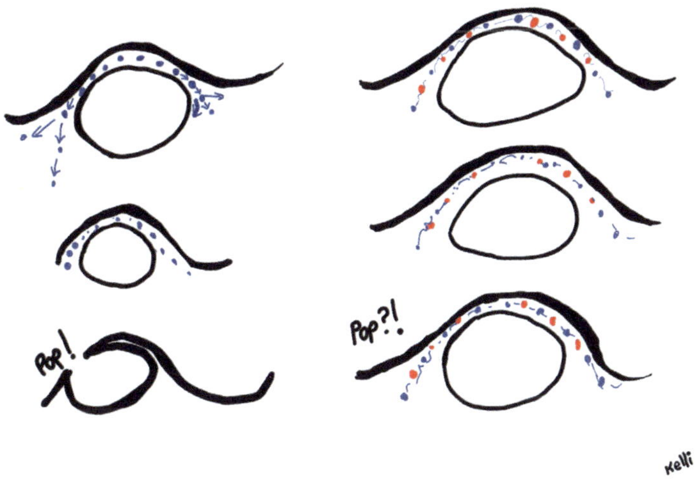

My physics problem was that it was impossible to include surfactants with the computational method I was using. By neglecting surfactants, I would not be able to predict how long the bubbles sat on the surface of the water.

To make matters worse, even if I could find a way to include surfactant effects, my length-scale problem reared its ugly head! For a single bubble, that thin film that controls when it will burst can be one hundred times smaller than the bubble itself—even more as it drains away. For that selfie on the beach, our camera now would not only need to be able to see the grains of sand on the shore; it would need to see the individual molecules that make up the sand as well.

"We have a physics problem." I slunk back into my advisor's office defeated (again).

"Okay. Let's talk it out," he said (again).

I explained to him that in real breaking waves, surface bubbles persist for a while, and in my computer simulations they come up to the surface and burst instantly.

I explained to him everything that I had learned in digging through the piles (and piles) of papers on the topic of bubbles and surfactants.

I explained to him that the ocean is full of surfactants—salt, chemicals, pollution, and even microorganisms.

I explained to him that I had researched my way into a corner and was unable to find a way forward.

I explained to him that not including surfactant effects in my research on ocean waves felt like leaving out another huge piece of our proverbial puzzle.

This was a long conversation, and I'm sure little of the scientific information I shared was likely news to my very patient advisor.

"What I want to know is," he said, "how important is bubble surface lifetime for what we want to study right now?" He firmly tapped the table between us when he said the words "right now."

I sat back, a little stunned. "I don't know."

"Maybe answer that question first."

Many conversations later, and many involving him saying, "This is what research is," I finally agreed to add surfactant physics to my list of future research topics so I could finish my PhD. It sat right below micrometer-sized bubbles.

I realized that in my eagerness to use CFD to understand how breaking waves affect air-sea interactions, I had given myself the *impossible* mission statement. There wasn't a computer on Earth[1] that could simulate all of the scales and the physics that we needed in a single simulation.[2] To become

1. This is not a hyperbole.

2. Twenty years post-PhD, there still isn't.

a better researcher, I had to find the humility to create possible mission statements.

I am a girl who wants to explore everything.

I devoured books about girls that flew on the backs of dragons. I followed creeks and paths in the woods behind my house just to see where they went. If there was some route home from school I hadn't explored, I took it even if I got home later than my mom thought was reasonable.

I am also a girl who asks questions (constantly) and likes to solve problems (especially hard ones). Taking that all together, what else could I be but some kind of a researcher?

This is how the ocean changed me.

I started as an aerospace engineer who worked at NASA and wanted to explore the moon. And then, on a fateful trip to Galveston Beach, I met the Thing that upended my life plans.

On my fateful dive trip to the Cayman Islands, it continued to upend my plans by making me rethink how I would be a ~~scuba-diving~~, ocean-engineering researcher.

Throughout my PhD studies, that Thing continued to challenge my ideas on what research is like and the importance of having "possible mission statements."

Twenty years after finishing my PhD, sitting at my desk on firmly, unmoving ground, staring at data and simulations on my monitor, I still think back to when I met the Thing on my first beach trip. The ocean, in all its vastness and unpredictability, has evolved from Thing to a beloved teacher. It reminds me of the balance between ambition and acceptance, exploration and humility. Every wave, every bubble, every problem I solve brings me closer to understanding not just the sea but also myself. And maybe, just maybe, that's the real reason I was drawn to it in the first place.

controlling eddies

The Story of a Girl Who Was Curious about Taming Turbulence

BEVERLEY MCKEON

Leaning my head against the side of an airplane flying over the Pacific Ocean during a family trip to Australia, I stared out the window at the clouds, watching the towering, bulging monsters and the small powder-puff stepping stones between them. I felt like I could plot my path through the clouds. I imagined I would be able to scale the side of the biggest cloud, which was taller than a skyscraper, using the bulges on its sides like climbing wall handholds, then hopping over to the neighboring one and . . .

Suddenly, the plane plummeted!

The person walking down the airplane aisle lurched onto the seated passenger in the row ahead, my juice spilled on my feet, and my book flew up and off the tiny seatback table. Passengers screamed and hyperventilated! What was happening?

Just as soon as it had started, we were back to the usual, steady flight.

The captain reassured us that we had run into an unexpected patch of turbulence, but that we had passed it and all was well. We were lucky. No one had been hurt, just a few folks shaken up. But it had made a big impact on me.

What *was* turbulence? Was it linked to the wonderful clouds I had been watching, and did I encounter it in other aspects of my life? Its effect on the plane was so dramatic; could we have avoided it? I was hooked, fully invested in understanding the phenomenon.

These questions stayed with me as I worked my way through to high school, but they weren't completely answered in physics or math classes. The subject was fascinating. I needed more information! So I majored in aeronautical engineering in college and learned a ton about the dynamics of fluids. Fluid: a material that flows; mechanics: the nuts and bolts of how that flow flows; dynamics: how to describe the motion of that fluid and how it evolves.

As you are reading this, there are flows of different kinds all around you. You are breathing: The flow of air in and out of your nose and throat moves along a tube that branches and forks to very small features in your lungs, where your body absorbs oxygen and desorbs carbon dioxide. Your digestive track is moving even trickier substances around too—a result of ingesting both solid food and fluids, chewing while your mouth adds saliva, swallowing while your esophagus adds lubrication, then digesting as your stomach mixes acid with food and absorbs nutrients. I'm not going to add vomit to

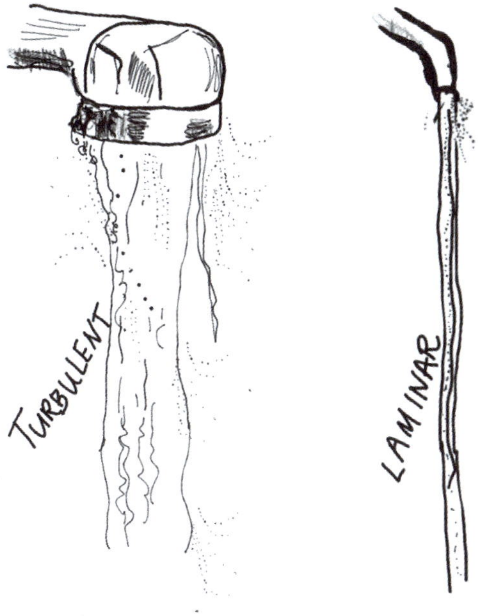

this description, but you get the idea that flows can carry "stuff" around!

Turbulent flows are ones which eddy, or rotate, and whirl around in an unsteady way. You can see the difference between turbulent and laminar (smoother) flows when you slowly turn on a faucet: The stream of water changes from smooth and steady—laminar—at low-flow rates to unsteady and noisy as it becomes turbulent when you hurry things up by opening the valve further. Flows we care about when we think of traveling somewhere ourselves or transporting utilities like water or natural gas around the world tend to be turbulent because of the speeds of travel and/or the size of the vehicle or pipe.

Now, back to my story. You may have heard of a movie called *Top Gun*. I saw it as I was starting to learn about fluid dynamics in the classroom and decided that I wanted to experience the phenomenon in practice by flying airplanes (fast ones!). The idea of being able to spend my days up in the clouds was very appealing. It scratched my itch to know more about the subject of turbulence . . . by experiencing being right in it!

However, as I learned more about flight and the field of aerodynamics in college—and got closer and closer to actually flying a plane myself—I realized that turbulence is a limiting factor in how fast airplanes can go and in how much fuel they have to burn to do it. Bummer!

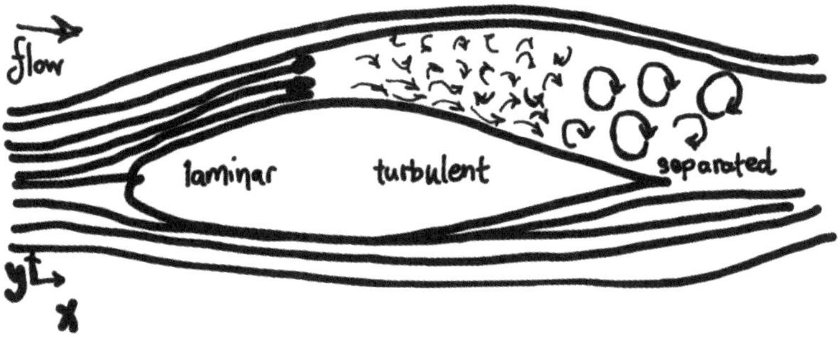

Turbulence in a thin layer right next to an aircraft wing can be responsible for half the drag or the force trying to hold an aircraft back, which must be countered by the thrust provided by the engines. This drag from turbulence in the so-called boundary layer is one of the reasons that so much aviation fuel is burned to transport an aircraft. There is also a boundary layer over the Earth, which controls the climate and in which turbulence transports the jet engine exhaust and other pollutants around in the atmosphere.

A light bulb went off in my mind: Rather than just *experiencing* and *trying to understand* turbulence, what about *controlling* it?

CURIOUS GIRL NOTE—What is turbulence, and why does it make plane rides bumpy?

- Turbulent flow: rotating, swirling flow that moves in an unsteady, noisy way.
- Laminar flow: smooth, steady flow.
- A thin layer of turbulence around a wing can account for *half* of the drag on an aircraft, which slows down the plane and uses more gas!
- Turbulence limits how fast planes can go and how much fuel they need.

I changed my mind again, distracted now by these scientific and engineering questions and the possibility of being able to do something that hadn't been done before. Could we make the air (or water, or gas) behave in a way that we dictated and allows aircraft to do new things . . . all without being able to see how the air is moving?! We needed experiments! Even before computing, humans used experiments and field observations to try to understand flow and turbulence. In a series of famous sketches in his diaries, Leonardo da Vinci drew the turbulent eddies and vortices he had observed by watching flow around and over objects in rivers and pools.

Making turbulence measurements under different flow conditions helps us understand what makes it tick. Humans have been able to write down equations for a century that tell us how both laminar and turbulent flows behave, usually in terms of forces and accelerations, just as Newton wrote. But we can't always solve them by hand; in fact, we can only do that for some of the simplest flows.

If the flow path is complicated—for example, around a winding geometry or covering a wide range of speeds—even the world's largest supercomputers can't simply spit out ways to either reduce the impact of turbulence in flows around objects like air or sea vehicles or exploit turbulence to give them new capabilities. Those questions are just too hard, because the behavior of each turbulent eddy—a tiny, identifiable vortex that whirls around in the soup of turbulence—is linked to and controlled by a host of others, bigger and smaller, slower and faster. To find out what makes turbulence tick, you have to understand the whole family of them and how they communicate with and influence each other. It's a bit like a turbulent game of telephone. One benefit of laboratory experiments is that the information is all there. We *just* have to be able to measure it all—and all at once! And then the *really* hard part is to make sense of all the information encoded in the measurements.

As part of my quest to control turbulence, I have been lucky enough to make measurements in the Earth's surface layer over the salt flats in Utah and on a radio-controlled car racing down a one kilometer track. The more I measured, the more I understood that new tools were needed to first measure everything and then to model turbulence. I needed to be able to *predict* what a given flow would do next so that I stood a chance of controlling it.

I realized that we needed to measure and model in space *and* time (instead of one or the other). Space-time allows us to observe how a flow evolves. We use these observations to try to understand how to make the equations simpler, to ask how much do we actually need to know in order to replicate and understand the key patterns and features of turbulence.

With a colleague, I started to design a model that tried to answer just this question using tools from fluid dynamics, linear algebra, data compression, and the theory of how to control linear systems. Reading books and papers and having to learn these different techniques made me feel like I was back to school. But it was a powerful feeling to bring them together and code up a model that might be able to unlock the secret of turbulence.

I pressed the "Enter" key to run the model for the first time and sat back, waiting for the computer to spit out an answer.

I am not a patient person and kept checking to see if the code worked.

. . . Did it?

I held my breath.

Nope.

At least we knew the answer quickly. (A feature of the model is that it is fast.) The numbers that should have been negative were positive.

What had gone wrong?

Fortunately, it was just a coding bug, and the numbers were supposed to be positive!

We had a working model!

Even better, our so-called resolvent model *did* allow us to identify what we needed to know to understand what we saw in turbulent flows as well as what kept the turbulence turbulent.

I would love to be able to tell you that we used the model to predict turbulent flow, revolutionize aircraft design, and tame turbulence . . . but you haven't seen that in the news, have you? The reason is that this is still an *extremely* difficult problem. Our model is one tool that has had some success in understanding turbulence but still needs more work before we can change the world. We made progress toward our goal, though!

There is another challenge before we can make turbulence behave as we want. Just as it can be hard to see flow, it is hard to change its behavior. I can push a pendulum and change its motion, but how do I get a hold of air? It's a bit like trying to shape a handful of Jell-O! We think up and test methods that give us a lot of "bang for our buck" (i.e., just tickle the flow without using much energy, but lead to a big change in the flow as a result). The tickling must usually be done at the surface so that any required piping, wires, and electrical connections can be simply made in several places—and repeatedly. Gotta keep that turbulence laughing!

You can probably think of several ways to do this: blowing air, moving little flaps or pistons, or even moving pieces of the wall up and down or side to side. I imagined an egg carton, where the ridges and troughs can change height in time but made it small enough to bump around the teeny-tiny, smallest turbulent eddies, with diameters smaller than the diameter of a human hair. I call this a *morphing surface*, with motion that you could probably feel with your fingertips but not see without magnification.

Ever impatient, I wanted to implement this idea but found it was not simple.

It took a detailed design process, state-of-the-art fabrication techniques, and some difficult wind tunnel experiments, but now we are able to use various types of morphing surface to change the behavior of turbulence. Harder still, we attempt to use our resolvent model to predict how the flow will respond and try to optimize how much of the turbulence we can modify. But most difficult of all is changing the flow as a whole in a beneficial way. This turns out to be much harder than changing individual eddies in a sea of millions of eddies.

Relatively recent technological developments also allow us to consider microelectromechanical systems (MEMs), which are arrays of microscopic ticklers, smart materials that can change their shape or texture on demand, metamaterials that are 3D printed to contain tiny beams and connectors in a regular pattern, or soft surfaces inspired by dolphin skins or bird feathers. There are many questions remaining to be tackled, ranging from how to design the ticklers to where and when to tickle as turbulent eddies rotate above them.

CURIOUS GIRL NOTE—How can we tame turbulence?

- Instead of experiencing and trying to understand turbulence, let's control it!
- How do we study turbulence? We need to measure and model in space *and* time!
- Laboratory experiments: wind tunnels, morphing surfaces, MEMs, and more.
- Computational models: equations too complicated to solve by hand so we need to code and run models.

So what had happened when the airplane had "hit turbulence"? And could I "fix" it with all that I had learned so far?

In an otherwise gently turbulent background of air, we had run into a patch of eddying flow so strong that it had tossed even our large passenger airplane around. The pilots were skilled enough to regain control and the plane big enough to stabilize, but things could have been worse if I had been in a small aircraft. Radar systems hadn't shown the dangerous turbulence, and the pilots hadn't seen it out the window.

Can I tame turbulence? Will I ever be able to? I can certainly change it, but not yet in a way that is always guaranteed to improve things! I have a model that makes turbulence easier to understand in terms of what we measure and what keeps the flow turbulent, and morphing surfaces that can tickle the flow into a new arrangement of turbulent eddies. But the big challenge—being able to force turbulence to do what I want—is still out there, waiting to be conquered. We still need to know more about this phenomenon. Still working on it!

Scientists started with inspired guesswork on when and where to push on turbulence to change it. Now, we are developing theoretical and computational models to predict how to get the changes that we want, at least in a limited way. These are baby steps. We are still working to master this extremely challenging and important problem. You don't often see turbulence taming devices in everyday life yet.

Success in forcing flow to be less turbulent could mean reducing the amount of fuel required to transport people or things around the globe, reducing air pollution. It could make turbines quieter. It could reduce dangerous patches of turbulence in cities where tall buildings can strengthen turbulent eddies. When we are successful, we will be able to influence how turbulence impacts the quality of life for people around the world.

CURIOUS GIRL NOTE—Why do we want to control turbulence?

- Save fuel and reduce air pollution
- Reduce noise from turbines
- Make cities with tall buildings that make dangerous patches of turbulence safer
- To do this, we need to measure, model, and test turbulence.

Little did I know as I sat on that bumpy flight that the turbulence we had hit and the swirling flows in the fluffy clouds I had imagined skipping across were part of a puzzle at the edge of what we know about the physics of the world. And that trying to solve this puzzle would turn me into a scientist.

finding engineering in every stroke

The Story of a Girl Who Loves Swimming

LEAH MENDELSON

The school swimming pool was gleaming and brand new, the most exciting news our small town had seen in many years, and one of the nicest within several hours' drive. The warm, humid air fogged my glasses the moment I walked in. Even through the fog, the sparkling blue water and bright white tiles immediately overshadowed the winter gray peeking through the windows. From the moment the pool was built, I lived there. As a teenager, you could find me there five or six hours a day.

At first, I loved the high dive, working up the nerve to do a flip. It was also easy to enjoy the days when our swim team would kayak, like clumsy bumper cars in the crowded pool, or play games. Our favorites were versions of water polo or sharks and minnows, but with fewer rules. Once we even turned the stretching-area mats into a slip 'n slide. After sliding along the length of the pool deck, I would land with a flop in the water.

Beyond these moments of joy and play, I also showed up for the training. Going faster was a quantifiable achievement made through hard work, and I wanted to succeed.

Unlike the buzz of a team on land, I cherished the mix of quiet and social time. Each lap in the water was a chance to think, daydream, or have songs stuck in my head. Our rest breaks at the pool's edge between laps filled the space with laughter and conversation.

It could take a full workout to finish telling a story. We'd share a few sentences at a time and then push off the wall to swim again, leaving each other hanging in suspense. I didn't mind that my friends would always ask me, the "good at math" one, to count laps or determine our next intervals. I knew we were teammates and in it together.

When you spend so much time on something, curiosity is a natural consequence. Swimming was a place for experiments. In practice, we would test small changes in our form that could make us faster. One day we would focus on the angle of our hand entering the water. Another day might look at when to bend our elbows, trying to maximize how far we moved forward on each stroke.

Sometimes there was homework.

"Stick your hand out the car window," a coach had suggested, "but look out for any signs or mailboxes. When it's clear, turn your wrist and feel how the forces change. How should you move your hand?"

Sometimes there wasn't a clear answer. I don't know if we ever settled our multiyear debate on whether it was better to put your arms directly over your ears or slightly behind them when streamlining, stretching out with our arms pointed and gliding for a moment after pushing off the walls.

Swim practice was also the first time I saw an underwater

camera. I didn't realize anything electronic could work underwater, let alone a video camera. I had always thought of photos and videos as a way to preserve memories, but this camera served another purpose.

The underwater camera was a tool that could help answer the questions we'd been asking about our swimming. After everyone had swum a lap alongside it, we gathered around an old TV to watch the videos and listen to our coaches' observations. Conversation ignited as we all invited each other to share what we noticed about our strokes.

Outside the pool, the advice focused on a different topic.

"You should be an engineer," my teachers and mentors pointed out. "You like math and science and doing things with them."

This was true, but not exclusive. I was arguably just as interested in music, books, and the internet (not to mention swimming). True, I was in the school preengineering class, but I dreaded it. This was the one place where I was constantly reminded that I was different—because I was the only girl in the room.

"Look what we made you," a boy's voice called from behind me one day in class. I turned around in my chair, with my hair dripping and wearing a T-shirt that said CHLORINE: BREAKFAST OF CHAMPIONS. His computer screen was filled by a cartoonish heart with my name in the middle, rendered in bright red in the computer's 3D drawing software.

There was always a joking proclamation of love or teasing of whoever I partnered with on class projects: "Are you two gonna go on a date after this?"

My classmates competed to see who could make the space the most uncomfortable for me. Most days, I would leave class either burning with anger or almost crying. It would take a good part of my afternoon's swim practice to shake off those feelings.

Even with what felt like everyone telling me I should be an engineer, I couldn't tell you what engineers actually did. I knew that we weren't talking about the kind of engineer that drove trains, but that was about it.

Every day in preengineering class, we would draw random shapes on the computer. They might as well have been a cat toy or a work of abstract art. Other than teaching us a new button to click on, they didn't seem to have any particular purpose.

From engineers I'd met, mostly through career days at school, I slowly pieced together that they could build cars or robots or helicopters or buildings or bridges. From TV I knew that maybe, if an engineer was lucky enough, they might get to design roller coasters. Roller coasters were pretty fun, so I decided that was my goal. I was going to be a roller coaster engineer.

In pursuit of my new goal, I chose to study at a small engineering college advertised around community and collaboration. I was excited about this camaraderie. Finally, there would be more than one girl in my engineering classes. Maybe I would even learn what all the random shapes I drew in my preengineering classes were for. But there were no sports, and definitely no swim team; the school was too small. I hoped that finding a community was worth it.

The small engineering school's classes were hard. In orientation, in between tours and trying out the cafeteria frozen yogurt, there was a pop quiz. I found myself sitting behind a classmate, watching his pencil flying while he wrote out all the physics he remembered from high school. I stared at my own blank page, wishing it could have been full of things, but I was pretty sure I'd never seen any of it before. I felt behind before I'd even started.

To top it off, my classmates seemed to want to spend their free time doing more engineering: building a robot in a research lab, joining the race-car design team, or teaching engineering to younger kids. Instead, you could probably still find me at the swimming pool.

Despite being darker, older, and more crowded than at home, the pool I found was comfortable and familiar. I planned and strategized my day around any short windows where the pool was open. While we didn't have a swim team, a neighboring school would sometimes let me practice with them.

I also picked up a job at a nearby gym as a lifeguard. I originally planned to use that pool, too, but found my efforts limited by its overheated bathwater temperature. My lifeguarding job became a place where I could just sit, smelling chlorine, surrounded by people who weren't already done with next week's homework.

The day things changed, I was sleep deprived in a math class, my hair still wet from a morning swim and once again wearing my chlorine T-shirt. The whiteboard was filled with endless rows of symbols and equations.

Highlighting a few of the lines, the professor explained that we can use this math to describe how a fluid, like water, moves around a cylinder. He sketched streamlines, contours of motion bending around a circle, on the board to illustrate.

I'd seen those before—in car commercials like the professor mentioned in his explanation, and also in swimming magazines.

Wait, my mind raced. *If flow around a cylinder can be explained by engineering, then why not flow around a person? Or the flow produced by a person?* Could minimizing splashing and air bubbles as my hand entered the water be engineering? Could changing the angle of your arm to maximize forces pulling through the water be engineering? Could deciding which laps of the long-distance race to cruise along at a comfortable pace and which ones to push faster be engineering?

As the professor continued his explanation of the equations, I grew more convinced. Problem-solving and experiments in the water could be engineering.

Excited by this realization and buoyed by the idea that I might finally answer the question about where to put your arms when streamlining off the wall, I signed up for another class focused on what I'd learned engineers called fluid mechanics. The homework wasn't easy, but to my surprise, the class went okay.

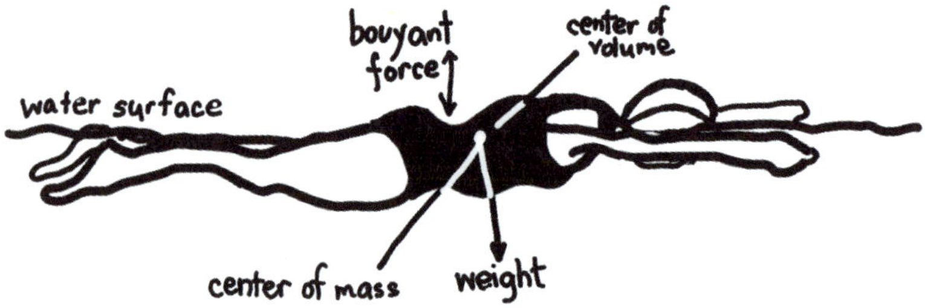

Whenever we needed to find an example of what we were learning about, I could usually turn to swimming.

Pressure was why your ears hurt at the bottom of the pool.

Turbulence was when you kicked really hard, creating a choppy mess of bubbles and water droplets moving in all directions.

Drag was when we would layer several baggy, worn-out swimsuits over each other to add extra resistance during practice, slowing us down like a parachute through the water.

Even after the class, I still didn't understand the flow around a person moving through the water. We had mostly learned about simpler shapes and motions like the cylinder.

CURIOUS GIRL NOTE—What I have learned about fluid mechanics of swimming:

- Water pressure can be large (what makes my ears hurt a long way underwater).
- Turbulence is caused by energetic kicking/choppy swimming.
- Drag is resistance to me swimming (like when we wear baggy swimsuits to make ourselves work harder in the pool).
- But how can I swim faster?!

To keep going, I applied to a research internship looking at how sharkskin could reduce drag, the force that resists a body moving through a fluid. I already knew about sharkskin and its unique texture of scales and ridges from the racing swimsuits the Olympians wore, like the one I'd once saved up my lifeguarding paychecks to buy before a big meet.

From my fluid mechanics class, I'd learned that, in addition to making swimmers faster, reducing drag could also make airplanes, cars, and watercrafts more efficient. It was time to see what I could contribute.

The water tunnel lab where I worked reminded me of my college swimming pool: dark, humid, and noisy. The windowless room was almost completely occupied by two large tanks that were connected by a windowed, rectangular box. Whirring pumps drove the water from one tank to the other, flowing through the windowed box before returning to the start through a maze of plumbing. It was hard not to feel like a real engineer while surrounded by all this equipment.

Working in the water tunnel lab was enthralling for another reason as well: The lab group worked as a team. I wasn't the only girl on the team, either! We each had our own research questions about the flow to ask and answer, but we set up and ran all our experiments together. With all the equipment to adjust and keep track of, we needed each other.

"Everyone good?" the other research intern called out, looking around to check that we all had our laser goggles on. "Laser on," he continued, flipping the power switch.

We needed to use laser light, which scattered brightly off the tiny glass beads that we put in the water, to trace how the water moved. Like in my old swimming pool, we used a video camera to film the particle motions. Even though this camera didn't need to be underwater, it often took an entire day to move, align, and realign the mirrors and lenses that bent and aimed the laser beam into the right shape. Then we still needed to place and line up the camera to see it.

"Okay, recording!" I hit "Go" on the computer and watched the screen timer count down. For each experiment, we needed to record video for a certain amount of time. "We're done," I announced as the timer reached the end.

"Laser off," my teammate followed. Like swimming laps, there were breaks in between experiment runs as we waited for our files to save. We could remove our protective goggles for a bit, blink, and relax.

"So why did you join this project?" He jumped back to our earlier conversation. For him, it had been airplanes.

"I majored in aerospace because it's rocket science," one graduate student weighed in. "But this project has a twist," she continued. "We're explaining how nature works and using nature to be better engineers."

"That part's important," I agreed. "I don't think I would have applied if not for swimming . . . Okay, we're done saving. Ready for the next run?" I flipped my goggles back into place. Our conversation would have to wait for the next break.

After we had the videos, we loaded them into a specialized computer program that would run for hours, calculating how fast and in what directions the laser-lit beads moved. While the computer chewed on our videos, you could find me either at the swimming pool or at the library, reading.

In my reading, I learned that sharks weren't the only example swimmer engineers were interested in; researchers were designing entire underwater vehicles that resembled and moved like fish, jellyfish, and even penguins. Swimming had a clear place in engineering.

As I scrolled through articles, I was surrounded by other students working on their research projects. After a few weeks, I realized I stopped keeping track of when I was the only girl in the room. Sometimes I still was, but this time there were no 3D drawings of hearts.

"Check out this video of my experiment I just took," another student shared, pointing to a computer screen showing cascading swirls of colorful dyes emerging from a nozzle. "We're going to lunch later," he continued, gesturing to the other students sitting nearby, "Want to come?"

Maybe I had a place in engineering too.

The more I worked on research, the more I realized that no swimming problem was just a swimming problem. Building the sharkskin model required mechanical design. To see our data, I needed to code. Analyzing results meant learning new physics.

I needed all the parts of engineering that scared me. The parts of engineering where I felt behind. But my classmates weren't there, and they weren't working on swimming. So like swim practice, I showed up for the challenge. I did the work for my team.

I'm now more comfortable with these uncomfortable leaps. But no matter how many new engineering skills I learn, swimming is still one of my tools.

I've moved on to studying fish and how they leap out of the water. At the end of a day in the lab, I still often head to the pool. I grab my fins and practice kicks—first slow and sweeping, then quick and small like the fish.

The water is where my curiosity started, and it's still where I experiment.

chasing the wind

The Story of a Girl Who Loves Sports

SHILPA VIJAY

I am obsessed with the sport of cricket. In India, cricket is more than a sport; it's a way of life. For those that don't know, cricket is similar to an extended version of baseball. It is a bat-and-ball sport in which teams take turns scoring runs by hitting the ball and defending wickets.

My family and I would huddle around the TV, cheering for our favorite teams. But what fascinated me the most was how bowlers could make the ball swing and deceive the batsmen.

I remember experimenting with different ways to deliver the ball.

If I used a certain angle from my wrist, I could outsmart the batsman. If I spun the ball while releasing it, it would take a completely different path. I was curious. Why did this happen?

People knew how to achieve the desired outcome, and TV experts would debate strategies endlessly. But when I asked people around me, no one could explain why the ball swung the way it did. I wanted to stay involved in cricket and figure out the mystery myself.

However, as I grew up, I realized there was no formal way for me to play cricket competitively because women's cricket wasn't as organized back then. I was hugely disappointed. But I didn't give up on sports. I played every sport I could—badminton, football, basketball. And with every game, new questions popped up in my mind.

Why is a badminton shuttlecock shaped the way it is?

Why does a football curve when you kick it just right?

Why do we throw a basketball in a certain way?

It was all about how these balls moved through the air.

I realized that science might help me find the answers. I also realized that to be taken seriously in conversations about sports—especially as a girl—I needed to understand the science behind the sport.

As I graduated high school, I was awarded the "Best Sportsperson," and my prize, ironically, was a book called *103 Scientific Discoveries*. *What's left to discover?* I wondered, staring at my new book. *It feels like everything useful has already been invented!*

As I flipped through the pages, I realized that scientists were still making discoveries every day. I discovered something fascinating called aerodynamics—the science of how things move through air.

Just like that, I had a renewed motivation to explore the science behind all of my beloved sports and answer the question, "How do things move in the air?" This book was a game changer for me.

While learning about aerodynamics, I discovered Kalpana Chawla's story—the first Indian woman to become a NASA astronaut. She used her knowledge of how things move through air to help fly spacecraft! Though she tragically died in the Columbia shuttle accident, her story showed me something amazing: The same science that made my cricket ball swing through the air could help rockets soar into space.

Even though India didn't have astronauts at the time, I couldn't stop thinking about the connection between my backyard sports and space exploration. Whether it was a curving cricket ball, a floating shuttlecock, or a spinning football (soccer ball, if you are in the United States), I was beginning to understand the invisible forces that made things fly.

Remember my burning question about why cricket balls curved through the air? The answer was right there in my hands all along! Every cricket player knows to keep one side of the ball shiny and smooth while letting the other side stay rough, but I discovered this wasn't just tradition. It was science.

Picture this: When you throw the ball, air races around both sides. On the smooth side, it flows fast and steady, like water from a tap. But on the rough side? The air tumbles and swirls, moving more slowly.

This difference in air speed creates a force that pulls the ball toward the rough side, making it swing! That's why bowlers spend so much time polishing one side of the ball on their cricket whites—they're actually setting up the perfect conditions for swing.

The faster you bowl, the more dramatic the swing. It's like when you stick your hand out a car window. At low speeds, you barely feel the air, but at high speeds, you can feel it pushing strongly against your palm.

Fast bowlers can make the ball swing so sharply that even the best batsmen get caught off guard! Baseball pitchers use

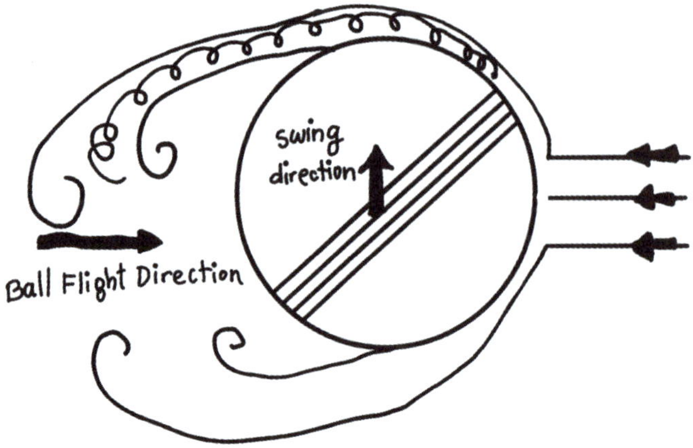

the same science. They roughen one side of the ball with their grip to make those amazing curveballs that leave batters swinging at air.

CURIOUS GIRL NOTE—Why does a cricket ball curve through the air when I throw it?

- Bowlers make one side of the ball smooth and one rough.
- Air flows fast and steadily on the smooth side, but it tumbles and swirls on the rough side.
- Difference in flow speed → force toward the rough side.
- The ball swings (moves sideways)!

Even my basketball shots made more sense now.

When you shoot a basketball, you want it to arc high and drop straight down through the hoop. This is where that backspin your coach teaches you becomes important. Backspin helps stabilize the ball's flight and creates a softer landing on the rim. Watch professional players closely. They all finish their shots with their hands pointing down, creating that perfect backspin.

Who would have thought that the same science that helps planes fly and rockets soar into space was right there in every game I played? These weren't just sports tricks anymore; they were real-world examples of aerodynamics in action.

An airplane wing is actually designed to make air flow faster over its curved top surface than its flatter bottom surface. Just like my cricket ball, this difference in air speed creates different pressures—lower pressure above the wing, and higher pressure below it. This pressure difference creates lift, pushing the plane upward against gravity.

When I learned that airplane wings work using the same principles as my spinning cricket ball, a light bulb went off in my head. Those endless hours in the backyard practicing my bowling weren't just about getting better at cricket. I was unknowingly running experiments in my own aerodynamics laboratory! Every serve, every kick, every throw was teaching me about the invisible forces that help planes soar through the sky and rockets break free from Earth's atmosphere.

Nature is full of amazing secrets about how things move through air and water, and sharks are one of the fiercest examples! Have you ever seen a sharkskin? Their skin feels rough, almost like sandpaper. But under a microscope, you'll see something incredible: Their skin is covered in tiny V-shaped scales called denticles. These aren't just for protection; they're nature's way of helping sharks swim super fast!

The way birds fly is another fascinating puzzle. Unlike airplane wings that stay still, birds flap their wings to create lift and thrust at the same time. Watch a hummingbird hover near a flower. Its wings move in a figure-eight pattern, pushing air in ways that let it float in place or even fly backward! When I learned this, I couldn't help but think about my cricket ball's swing. It's all about controlling the invisible movements of air.

Think of it like this: When you run, the air pushes against you, slowing you down. The same thing happens to sharks in water and planes in air. This pushing force is called drag. But nature is smart. Those tiny scales on their skin guide the water in a specific way that reduce this drag, letting them glide through the ocean like underwater rockets!

In my lab, we study all these incredible tricks from nature and try to apply them to real-world engineering problems. Using special cameras and wind tunnels, we can see how birds change their wing shapes during flight and how fish create helpful swirls in water. These discoveries help us design better race cars, more efficient airplane wings, and even underwater vehicles that move like fish. It's like being a detective solving mysteries hidden in the air and water around us. But instead of just solving the mystery, we get to use the answers to build amazing things!

Remember how my cricket ball curves through the air? That same science is helping us design better vehicles. By understanding how air flows differently over smooth and rough surfaces, engineers are creating new airplane surfaces that could make flying more efficient and use less fuel. Nature has already mastered these tricks. Shark scales and bird wings have been perfecting their designs for millions of years.

We see these principles at work in many engineering applications today. For example, Formula 1 cars use wings that push the car down instead of lifting it up, helping them grip the road at high speeds. Like birds adjusting their wings during flight, these cars can change their wing positions while racing using a system called DRS (drag reduction system).

Modern trucks have similar features. Those panels you see on their sides and top help reduce drag, saving fuel. Even high-speed trains are shaped to minimize air resistance, making them more energy efficient.

The science behind a cricket ball's swing connects to all sorts of real-world applications. From airplanes and cars to trains and ships, understanding how things move through air helps us build better, more efficient vehicles.

Not to be dramatic, but reading *103 Scientific Discoveries* truly changed my life. Without the gift of that book, it's possible that I would have never realized how many interesting problems were waiting to be solved. I had always thought of science as a collection of answers. But that book taught me that science is really a collection of questions—questions that demand creativity, persistence, and curiosity to explore.

Looking back, I realize how much sports and science are alike. Both involve observing patterns, testing limits, and learning from failure. In cricket, I experimented with wrist angles and grips to make the ball swing. In badminton, I watched how the shuttlecock danced through the air, defying my instincts about its movement. In basketball, I practiced spinning the ball just right to give it the best chance of arcing into the hoop. In science, my experiments aren't so different—I test, fail, learn, and try again.

Science and sports both thrive on curiosity and precision. They're about asking, "What if?" and then pushing forward, whether on the field or in the lab.

Sports sparked my questions, and science gave me the tools to find answers.

As a researcher, I know that even the smallest observations—like a cricket ball swinging through the air—can lead to breakthroughs that reshape how we understand the world. The same curiosity that made me wonder about the curve of a football now drives me to study airflow, design better surfaces, and explore the mysteries of fluid dynamics.

Both sports and science remind me that every answer sparks more questions, and every challenge holds the promise of discovery. Now when something soars overhead—a bird, a plane, or even just a thrown ball—I see both the magic and the science of flight. That's what sports taught me: Behind every graceful movement, there's a fascinating puzzle waiting to be solved.

I may have started by just wanting to bowl the perfect cricket ball, but I discovered something much bigger: the amazing science that connects everything from a spinning cricket ball to a spaceship breaking through Earth's atmosphere. And every time I watch something soar through the air now, I don't just see it move. I understand the invisible forces that make it fly.

wind, words, and wildfires[1]

The Story of a Girl Who Wants to Measure Things

RONI GOLDSHMID

As a fourth grader, I couldn't wait for the end-of-year party we had been planning for weeks. The excitement was all I could think about for days. My brother and I sat on the couch in the afternoon after school one day, lazily sharing a bag of chips. Suddenly, Mom walked in, balancing a tower of boxes in her arms. The sight was strange enough to start questioning the situation in front of me.

"Who are the boxes for?" I asked, curious but unbothered.

She set them down and looked at me, her face calm but serious. "We're moving to Los Angeles. Where we visited last

1. This piece was written before the devastating wildfires of 2025 in Los Angeles. The author continues to strive to create technologies to help firefighters during these kinds of events.

summer. Dad will go tomorrow, and we'll follow as soon as we find tenants to rent the house."

The words hung in the air, heavier than the boxes she carried. I sat frozen, the familiar comfort of the living room suddenly feeling less like home.

The news of our move to Los Angeles made everything else—the party, my friends, my extended family, my familiar life—feel like it was slipping away, growing distant and unreachable.

The day before we were set to leave, my classmates surprised me. They handed me a stack of goodbye letters, each one bursting with scribbled memories and heartfelt wishes. They even threw a tiny celebration right there in the classroom, with a cake and balloons. Their laughter and hugs wrapped around me like a warm blanket, offering a small flicker of comfort as I faced the unknown.

The biggest unknown and my biggest challenge as my family settled into Los Angeles was the language barrier. Understanding simple conversation was one thing, which I was not yet capable of either; but in subjects like science, history, and physics—it was something else. My school did their best to help, assigning me a personal tutor to guide me through the maze of English for several hours each day.

Two weeks into fifth grade, my teacher came up to me, apology written all over her face. I really couldn't understand her and needed the help of another classmate who had emigrated to the United States a year before me and could speak Hebrew and English.

She leaned over and translated, her voice soft but clear. "The tutor can only stay until 11:00 a.m. every day because of budget and schedule limitations."

I froze, the weight of the news sinking in. "Wait," I blurted, panic rising in my voice. "So for history, science, and math, I have to sit with the rest of the class? But I can't. I don't know English for this! Why can't she be gone just for math time?"

My words came out fast, almost desperate, but they were true. Math was a universal language, so I could work with that.

But how can I do history and science without an English tutor? The thought of facing the rest of the school day alone, without understanding a single word, felt impossible.

I found myself struggling every day to keep up with the pace of the class, often feeling lost and confused. Math has a universal language of numbers and symbols. Science uses English to explain the concepts—the double challenge learning a subject and a language at the same time! Each day after school, I came home in tears, collapsing into my mom's arms.

"I can't do it, *Ima*,"[2] I sobbed one afternoon. "I don't understand what anyone is saying. How am I supposed to learn science and history when I can't even read the questions?"

My mom knelt in front of me, holding my shoulders. "It's hard now, I know," she said, her voice calm but firm. "But kids learn new languages fast, you will see. Plus, you're brave. One day, this will all feel easy. You just need to keep trying, even when it feels impossible."

Her words didn't erase the frustration, but they gave a small spark of hope that my struggles would get easier someday.

I remember my mom explaining, "Being brave does not mean that you're not scared. It only means that you decide to act despite being scared."

Remembering that my mom called me brave, I immersed myself in the English language, spending countless hours studying, practicing, and seeking out opportunities to converse with native speakers.

I accompanied my parents on errands to banks and lawyer's offices, forcing myself to quickly improve my language skills by using them in practical, real-world situations.

By the time I was nine, I felt like an adult, shouldering responsibilities beyond my years, and my belief in my abilities

---

2. Ima means "mom" in Hebrew.

grew. This belief, however, sometimes felt like it was hiding in the shadows.

One summer afternoon before sixth grade, one of my mom's friends came to visit us all the way from Israel. She was a teacher, and I remember looking up to her and valuing her opinion. As we sat around the table eating watermelon, she turned to me with a thoughtful look.

"Now that you're going to middle school," she said, her voice almost casual, "you have to start thinking about what you want to do when you grow up. Did you decide yet?"

I hesitated, feeling the weight of her question. "Um, I don't know yet," I mumbled, avoiding her gaze.

"Well," she continued, leaning forward, "to be successful, you should earn a million dollars. Do you know how long that would take if you made a hundred dollars a day? Calculate it."

I paused, doing the math in my head. "That's ten thousand days," I said finally.

She nodded. "Right. Now, how many years is that?"

"Twenty-seven," I replied, my voice barely above a whisper.

"Exactly," she said firmly. "That's twenty-seven years of working every single day and saving every dollar without spending a penny. You see? It's nearly impossible. You have to go to school and get an education if you want to earn more money than your parents do now because they did not go to college."

I stayed silent, nodding along to her words. But inside, I felt a wave of fear and pressure, as if the weight of my entire future had been dropped on my small shoulders.

What she said sat on my mind for the rest of the summer. The only thing I knew for certain was that I only wanted to explore, discover, build things, and help my community. I also remember knowing that I would never be happy doing the same thing every single day for the rest of my life.

I immediately tried to think about what I was most interested in. I knew that growing up in water-scarce Israel had instilled in me a deep respect for conservation, which steered me toward the field of environmental engineering. But moving to Los Angeles lit a flame in me (pun intended). The more I heard about the devastating wildfires, the more I wanted to help conserve our wildlands, including trees and animals from the impact of those devastating wildfires. I yearned to understand these environmental challenges and to find solutions, but the language barrier continued to make me feel like I had a brick wall to climb every minute of the day.

Yet a chance encounter, a couple of years later, with a guidance counselor opened a new path. *This will finally be my opportunity to not feel behind for once*, I thought. My friends warned me that it would be a hard path to take. But I was stubborn and didn't listen. I knew, deep inside me, that it was my opportunity to break down this brick wall that'd been haunting me for years.

The counselor said, "How would you like to earn high school and college credit at the same time—for free?"

I was immediately intrigued.

She continued, "You can now enroll in community college courses. Each class counts as a full year of high school credit, fulfills general education requirements for college, and lightens your high school workload. It's faster and more affordable than AP classes or SAT prep."

What I learned from her in that moment was, *If you have the opportunity—explore it. You might discover something new about your interests and abilities.*

Her words revealed an option I hadn't considered as this was a chance to act toward getting an education that my parents never got, and I embraced it. I was scared, but I decided to be brave because finally, for once, I knew that I would be able to feel like I wasn't playing catch-up, but that I was ahead.

And I did it. I completed so many credit hours during high school, which allowed me to transfer to UC Berkeley as a junior straight out of high school. But even with that accomplishment, which clearly made me get ahead of many of my classmates, the feeling like I was behind found ways to show up at times. For example, as a first-generation college student, I hadn't even heard of fluid mechanics until my master's degree.

Most people don't know how cool the field of fluid mechanics is.

One of the professors in my master's degree program specialized in fluid mechanics. His specialization focused on the physics of ocean waves and the atmosphere, and his work immediately caught my interest. I joined his lab and focused my graduate research on how mountain winds develop from solar heating.

These flows are part of the atmospheric boundary layer—the lowest layer of the atmosphere, directly influenced by Earth's surface. This layer is critical for understanding wildfires, as it drives wind patterns, and it's where air pollution is most concentrated due to human activity.

Fast-forward, and I am now an aerospace engineer who builds devices that help us measure and understand our environment. I took a crazy and winding path to get here, but that is a story for another day. For now, just know that those recurring California wildfires had a big influence on what I do today.

Speaking of wildfires, one of the biggest challenges firefighters face today is the lack of accurate, real-time information about the wind. Believe it or not, when they need to measure the wind speed near a fire, they hold a handheld device called an anemometer, which looks like a small fan, up into the wind. As the blades spin, the device calculates the wind speed and displays it on a screen.

CURIOUS GIRL NOTE—How do the atmosphere and wind affect wildfires?

- Earth's atmospheric boundary layer: the lowest layer of the atmosphere, which is directly influenced by the surface of the Earth.
- Solar heating causes mountain winds.
- Important for wildfires because this drives wind patterns!
- Challenges with wildfires: lack of accurate, real-time info about the wind.

The firefighter holding the device then shouts the wind speed to nearby team members, and together they use this information to decide how to proceed. It's a simple method but can be problematic in the fast-changing wildfire environment, where every second counts and the wind can shift in an instant.

There are many ways to measure wind, but some of them can get pretty complicated. One tool, called a sonic anemometer, uses sound to measure how fast the wind is blowing. It sends out tiny sound waves—way too high-pitched for us to hear—back and forth. If the wind is blowing in the same direction as the sound, it helps the sound travel faster, like giving it a little push. If the wind is blowing against the sound, it slows it down. By carefully measuring how the speed of the sound changes, the anemometer can figure out how fast the wind is blowing and which direction it's coming from. These tools are very precise and can even measure gentle breezes!

But what if you want to understand how the wind moves in lots of different places at once, like in an open field somewhere in the mountains near a wildfire? Setting up these sound-measuring tools everywhere isn't very easy nor practical, and it's also very expensive. That's where another method comes in: visualizing the flow of air using particles.

Wind, like all moving air, is usually invisible. To "see" it, scientists use tiny particles that are so small and light—ideally their weight is the same as the air—they allow them to float along with the air and trace its motion. Imagine sprinkling glitter into the wind and watching how it moves . . . That's the idea! These particles act like markers, showing us where the air is going and how fast. A bright laser lights up the particles in one area, and a superfast camera films their movements. By looking at the video frame by frame, we can figure out the speed and direction of the air in that spot. This method is called Particle Image Velocimetry, or PIV for short.

But even PIV can be tricky to use in large areas, like forests or remote locations. It can also be a source of pollution, and we don't want to cause harm to the environment when we measure the wind. That's why scientists are coming up with creative new ideas to measure the wind.

One idea that I focus on is to use things that are already in the environment, like trees and grass. When the wind blows, trees sway and grass ripples. By studying videos of vegetation in the wind, we can learn and estimate how the wind is behaving—without needing fancy equipment. It's like letting nature itself helps us track the invisible wind.

CURIOUS GIRL NOTE—Tools to measure wind speeds:

- Anemometer: handheld device
- Sonic anemometer: uses sound (too high-pitched for humans to hear)
- Particle image velocimetry: using tiny particles that float in the air to "see" how the wind moves
- Idea! Can we use the motion of trees and grass in the wind . . .?

Have you ever just sat and looked at how trees sway in the wind? If not, give it a try. Pay attention to how the canopy or the branches and leaves sway on different days with different winds. After a few times of paying attention, I bet you would have a sense of whether the wind is weak or strong and maybe even which direction it is blowing.

But what if we wanted to measure it more precisely? That's exactly what some of my current research focuses on.

To study how trees move in the wind, I created some fun experiments! First, I watched real trees outside and took videos of them swaying. But that wasn't enough. I wanted to control the wind, so I used a giant outdoor wind tunnel. Imagine a large open space with a wall that has 1,300 fans that can create any kind of wind you want! I put big trees inside the wind tunnel and watched how they moved when I changed the wind speed. I also had two cameras recording their motion and a sonic anemometer to record what the actual wind was like so that I can study it all together later.

I also took my experiments to the desert, where I planted even larger trees in a large open space. I used a drone to film real trees while measuring the wind speed with the sonic anemometer. Flying a drone in windy conditions is tricky, but it helped me understand how the drone would see the trees' movements in real-world situations, just like firefighters would need to know.

After studying thousands of videos, I discovered something fascinating: Trees with big leaves sway much more dramatically than trees with small leaves, even when the wind is the same. This was my aha moment! It helped me develop a method to estimate wind speed from simple videos of trees swaying.

Want to try it yourself? The next time it's windy, find two trees that are close together. One should have big leaves, like a maple, and the other should have small leaves, like a pine or oak. Watch carefully as the wind blows. Which tree's leaves sways more? The one with the big leaves, or the one with the small leaves? You might be surprised to find that the tree with bigger leaves moves more, even though they experience the same wind! It's amazing how much we can learn from simple observations like this.

When I first trained a computer program, called a neural network, to estimate wind speeds from videos of trees swaying, it worked well on trees it had already seen. But whenever I added videos of new trees, the program needed calibration, like tuning an instrument to keep it accurate.

Calibration is essential in science to get accurate measurements, and here's an example:

Imagine you're trying to measure a tiny cricket that won't stay still. You decide to take a video of it, thinking you can measure it later. However, if you watch the video on a small phone screen and then on a big TV screen, the cricket will look like different sizes. To solve this problem, you can use a ruler. Place the ruler next to the cricket in the video. This way, no matter how big or small the screen is, the ruler and the cricket will always change size in the same way. By comparing the cricket's size to the ruler in the picture, you can accurately measure its real-world length, even if it's moving around! That's calibration.

CURIOUS GIRL NOTE—How can we measure wind speeds from tree motion?

- Take videos of different trees swaying at different speeds—but different trees have different leaf sizes that move differently!
- Train a neural net (artificial intelligence) to estimate wind speeds from these videos.
- Calibrate the neural network to measure accurate speeds, regardless of the tree type or leaf length.

In the same way, I needed to calibrate my program so it could work on videos of trees with different leaf shapes and motions. After weeks of effort, it finally started to make correct estimates. What I gained from this wasn't just a working program, but also skills I'll use in every challenge I face as a researcher. Designing experiments, analyzing data, and learning from mistakes taught me that engineering requires iterative learning, which is a process of doing and improving multiple times.

My journey through academia began as a simple curiosity about environmental issues eventually leading me to the fascinating world of fluid mechanics—a field where invisible forces shape everything from the flight of a mosquito to the power of a wildfire. Whether studying the aerodynamics of race cars or uncovering the secrets of wind patterns, I've found that fluid mechanics offers endless possibilities for discovery.

Overcoming challenges, like language barriers and academic struggles, has taught me resilience and the value of persistence. Through this journey, I've learned that challenges aren't obstacles—they're stepping stones. By embracing them, you may discover a career that's not only fulfilling but also impactful. I'm grateful for the opportunities I've had to design experiments, mentor students, and contribute to advancing scientific knowledge.

As Owen Cook once said, "Life should be about adding value and leaving a legacy, helping others, and having as much fun as humanly possible."

For me, fluid mechanics has given me all four.

So, yes, this is the story of a girl who dared to measure the invisible and found a world of possibilities waiting to be explored, while making sure she doesn't do the same thing every day.

facing strong headwinds

The Story of a Girl Curious about Wind Energy

KAREN ASHBY FLACK

Bang! The race starts. I am rounding the first curve of the 400-meter dash. Nervous energy propels me forward. I am fast, but not that fast. Mostly I enjoy running because I get to hang out with my friends. However, this time, I am really flying on the backstretch. Rounding the second curve to the final straightaway, I understand why I was breezing through the race: I ran into a wall of wind. I am not sure what place I got, but it felt like a victory just to make it to the finish line!

While the backstretch tailwind pushed me forward, the straightaway headwind stopped me in my tracks.

In my childhood hometown of Albuquerque, New Mexico, there are four seasons: summer, fall, winter, and *wind*. My memories include chasing dust devils on the dirt playground and running down the street after tumbling tumbleweeds, pigtails flying. All this motion and resistance to motion got me wondering, *How do you harness the energy in the wind?* Of course, people have been using the wind to propel sailboats across vast oceans and spin windmills to grind grain and pump water for centuries. But what if many things could be powered by these devices?

The wind approaching a windmill or wind turbine has kinetic energy, which is the energy of motion. The higher the wind speed, the more kinetic energy. The blades of the wind turbine, shaped like airplane wings, are attached to a central hub called the rotor. Wind blowing over the blades creates lift, which pushes on the blades so they rotate in a circle. The rotor is attached to a shaft that spins magnets around coils of conductive wire, generating electricity. This is a generator. The turbine converts the energy of the moving wind to mechanical energy in the spinning shaft, then to electrical energy using the generator.

CURIOUS GIRL NOTE—How do wind turbines work?

- Kinetic energy (KE): energy of motion
- More wind speed → more KE
- The wind creates lift on each turbine blade, which causes the wind turbine to rotate in a circle
- The spinning motion of a rotor around the shaft spins magnets around metal wire coils, which generates electricity
- KE of the wind → mechanical energy of the wind turbine shaft → electrical energy using a generator

It turns out that while I was running laps around the track in middle and high school, scientists in New Mexico had the same question. They were taking advantage of these breezy conditions, trying novel ways to capture the energy in the wind. I was able to see a few of these fascinating devices when my class took a field trip to Sandia National Labs. They looked something like the windmills I had seen on ranches and farms but taller and with only a few long blades. Others looked like big eggbeaters. A scientist enthusiastically told us that someday our homes and entire cities could be powered by these machines. Wind turbines . . . This could be the next big thing! I was hooked. How could I be a part of this?

The power (energy divided by time) a wind turbine can generate is very dependent on the diameter of the turbine rotor and the wind speed. If the diameter of the rotor doubles, the power quadruples (4x). If the wind speed doubles, the power can octuple (8x). That's why the turbines have long blades on tall towers in windy locations.

CURIOUS GIRL NOTE—What factors contribute to wind turbine power?

- Power = Energy/Time
- Power is related to the diameter of the rotor (2x diameter → 4x power!)
- Power is related to wind speed (2x wind speed → 8x power!)
- Rated power: actual power (measured in kW) generated by the wind turbine, based on the blades and generator

I learned if I wanted to be a part of this world, I should study mechanical engineering in college. The moment I entered college, I felt like I was lining up on the starting blocks again. Similar to my workouts in the foothills of New Mexico, I was starting a training run of discovery and effort. Maybe someday I could help design a wind turbine. Some classes I sailed through, but others . . . Well, let's just say that college can be hard, especially when you look around and hardly anybody looks like you. I was one of only a few women studying engineering. Also, I did poorly in my favorite subject, fluid dynamics. This was the class that taught the science of wind turbines. Not my best performance. I had been training hard, but obviously not hard enough. Maybe I didn't belong? Maybe I wasn't smart enough? Nope. I had faced strong headwinds before. I put my head down, pumped my arms, and powered through.

Wind turbines can't capture the energy in the wind perfectly, but they do a pretty good job. The actual power they generate, based on the size of the turbine blades and the generator, is called the rated power, and it is measured in kilowatts (kW). What's a kilowatt in everyday terms? Your cell phone uses a bit more than

1 watt (W), or 0.001 kW; your laptop uses about 100 W (0.1 kW); and your toaster needs about 1000 W (1 kW). The average home in the United States uses about 1000 kilowatt-hours (kWh) every month. If you used your toaster for an hour, that would be 1 kWh. A kW is a unit of power, and a kWh is power × time = energy.

Shortly after graduation, while moving from New Mexico to California, I crossed over the foothills into Silicon Valley and caught sight of an entire field of wind turbines spinning away. I pulled over to read the information sign. Before me was the world's first wind farm at Altamont Pass. Now I was definitely curious. How do these beautiful machines work? How could I work in this field?

The first modern turbines installed at Altamont Pass had blades 17 meters in diameter and a rated power of 100 kW. A turbine doesn't always produce that power since the wind is not always blowing fast enough. Taking that into account, a single one of these turbines generated enough power in a year for about twenty-five homes.

A lucky break blew my way. The best place to study aerodynamics was in a wind tunnel, and I lived by the world's largest wind tunnel at NASA Ames Research Center in Mountain View, California. This wind tunnel was so large that full scale planes could fit in the test section. Feeling a breeze at my back, I applied for a job and got it! I worked with NASA engineers, learning as much as possible about wind tunnel tests and aerodynamics. It was here I realized that I wanted to do experimental research, where I needed advanced degrees. More school? Why not? I had a goal, experience in the wind tunnels, and muscle memory facing headwinds.

Time to lace up the shoes for the marathon called graduate school. I warmed up with a few classes, attending part-time, and then committed to full time. I ran hard, not always fast, but I kept going. Thankfully, I had other students as my training partners. Along the way, there were uphill climbs (hard classes, qualifying exams) and downhill coast (fun lab mates, the joy of exploration). Even some unexpected hurdles, including a major earthquake that broke measurement equipment and caused damage to the laboratory. Sometimes I had good ideas, and sometimes I didn't.

Researchers face setbacks all the time. That is how you learn what works and what doesn't.

I truly enjoyed being in the laboratory, and I was becoming good at experimental research. After six years of graduate study, I earned a PhD in mechanical engineering, with a specialty in fluid dynamics. Honestly, most of it was fun, and along the way, I learned a whole lot more about how wind turbines work.

CURIOUS GIRL NOTE—Power and energy comparisons:

- Phones use 1 W
- Laptops use 100 W
- Toaster uses 1 kW (or 1000 W)
- Average US home uses 100 kWh (energy) every month
- First modern turbines at Altamont Pass had a rated power of 100 kW, but a turbine doesn't always produce that much because wind speeds can be slower. One turbine could power twenty-five homes in a year!
- Some wind turbines have a rated power of 5 megawatts (MW), or 5000 kW—enough for a single turbine to power over 1,000 homes each year!
- The biggest ocean turbines have a rated power of 15 MW, enough for 5,000 homes each year!

My next distance run was on a challenging and rewarding path. This was my career as a professor at the United States

Naval Academy. For nearly three decades, I taught future naval officers aerodynamics and hydrodynamics and performed many experiments in wind and water tunnels. I even taught a specialty class on wind turbines. The results of my research, from countless hours playing in the lab, helped scientists and engineers design the modern turbine—an engineering marvel.

Some wind turbines are 145 meters in diameter with a rated power of 5 MW (5000 kW). That means their blades span one-and-a-half football fields. Today, one of these turbines generates enough electricity for over 1,000 homes every year. Wind farms are also being built in the oceans near the coastline. Offshore turbines can be even bigger, since it is easier to transport a big object over the water on a barge than on roads and railways. The biggest ocean turbines have a rated power of 15 MW and rotor diameters over 220 meters—the length of two football fields, including the end zones. These machines are enormous. A single turbine like this could provide enough power for more than 5,000 homes in the United States every year. Wind is powering our homes, our cars, our phones. It happened!

Taking a good idea to an established solution doesn't just happen. It takes years of research, dedicated scientists, and curious people's dreams to make it possible. Someone has to do the work.

I wanted to be—and became—one of those people.

There is more work to do with new turbine designs, including floating turbines in the oceans and underwater turbines that harness energy from currents and tides. Who knows what else. Aren't you curious about the next big thing? I am.

Now as I head out for my morning workout, that curious young girl is still with me, pigtails flying. A headwind still slows us down, but it doesn't stop us in our tracks. It makes our morning coffee.

fluid mechanics to grandma's rescue!

The Story of a Girl Who Must Discover Tiny Labs

MARIE HÉBERT

As I go through the door, I call out, "Hello, Grandma, it's me. How are you doing?" I kiss her on the cheek, but I notice her worried eyes. The nurse has already arrived and is taking care of her.

Grandma whispers under her breath, "What is the name of the new nurse?"

I answer patiently for the nth time. "This is Riley, who helps us take good care of you."

She sighs in relief and relaxes.

Grandma also often asks about her next-door neighbor who moved out a few years ago. Soon, she will forget me, and I cannot let this happen. The doctors have exhausted all available treatments.

The last medication we tried made her vomit back everything. She could not even hold down a piece of her *sucre à la crème*. The previous medication did not do anything. And the one before, she started swelling like a balloon.

The doctors suggest focusing on managing the side effects. However, some labs are working on developing better medication for people suffering from Alzheimer's disease. This sickness makes Grandma's brain work differently, like when she forgets the nurse's name. Even if her cognitive decline cannot be stopped, my objective is at least to slow it down. Since our conventional treatment options ran out, I have been visiting each of the labs systematically.

The lab from last week is currently in its early testing phase. But each phase will last many months and potentially years!

The other conventional lab closed, which explained why my phone calls went unanswered. They could not afford the rent for so long. Their bulky equipment needed an expansive and expensive space. This eliminated one of my last options. Grandma's memory is deteriorating, and none of the medication works.

It is time for a less conventional solution: tiny labs.

I heard rumors of the Tiny Lab Association and its exclusivity.

Only people who follow their rigorous training have access to the network of tiny labs spread across the world. Their achievements are impressive. Tiny labs replace the big, clunky tabletop machines that require hours of hands-on effort. Tiny labs can analyze tiny droplets cheaper and using less space. Most importantly, the tiny labs can test much faster than the old-fashioned way. Tiny labs could be the key.

I get to the address of the Tiny Lab Association, but the door has an access card reader. I cannot go in. As I wait, I review the information folder I prepared. The Tiny Lab Association has strict rules. To use their tiny labs, you have to finish a special apprenticeship.

The association initially had less strict requirements. However, people without the proper training clogged, broke, and inefficiently used the equipment. They then restricted the membership.

The tiny lab experts continue their work across the globe. Their objectives focus on better understanding fluid flow at such a tiny scale and developing tools for many applications. However, the adoption of the tiny lab methods has not reached its full potential yet. Some rebel members of the Tiny Lab Association grant access to other scientists for collaborative projects. The work usually yields results despite the numerous obstacles. Communication between the Tiny Lab Association members and other scientists can be difficult due to the difference in background and expertise.

The exclusivity does not discourage me. The Tiny Lab Association represents my last hope. Their quick tests would enable the fast development of a medication. First, I must contact a researcher. I sit on a bench not too far from the locked door.

A woman hurriedly exits and closes the door behind her. I jump up and walk beside her. "Sorry, ma'am. Sorry, ma'am," I say. She stops and turns around. I begin explaining my situation to her.

She interrupts, "Sorry, I cannot help you. I specialize in microfluidics, not in the pharmaceutical field."

Her tone stuns me. I do not answer, and she walks away as I stand there, frozen. I feel the tears swell and my throat tighten. Her refusal felt like a scolding. Defeated, I drag my feet to the closest bus stop to go back home. It feels like the bus will never come, and I will not get back to Grandma in time.

By the time the bus arrives, I have breathed deeply, dried my tears, and recomposed myself. As I approach home, I get more and more energized. I am progressing in the right direction. I will refuse no as the final answer.

Instead of going back home, I stop at the library. I need a plan, and I am determined to learn everything I can about these tiny labs.

After many nights spent at the library, I finally better understand how tiny labs work. I already had a good base about fluids along with knowledge of automation and coding. However, I will need help to master the tiny lab equipment. I will also require a pharmaceutical expert to tweak the molecules just right. Now, my goal is clear: get an apprenticeship at the Tiny Lab Association.

The next day, I go back to the tiny lab's unassuming entrance. Determined to get what I need, I wait for someone to exit. This time, I am prepared with an elevator pitch and my CV.

The door finally opens, and a scientist exits the building for his break. Thankfully, I recognize him from my research: Brad. This researcher started a simple automation system of tiny labs to make them more accessible and repeatable. I take a deep breath in, filling my lungs with the fresh air, and a deep breath out, releasing all the tension from my body.

I grab his attention by using his name. I say hi, introduce myself, and ask him for an apprenticeship at the tiny lab facilities. "Here is a cover letter, CV, and references for you," I blurt as I hand him the documents. Brad confirms my initial understanding after I explain my goal. The tiny volume is the key to the fast reaction that will let me test many potential drugs quickly.

Brad concludes, "I have all the information I need. Let me think about it, and I will be in touch." Unsure whether I will hear back from him, I head home.

A few days later, Brad calls to invite me to join the Tiny Lab Association as an apprentice. This is great. I am simultaneously excited and stressed. I must keep my focus on what matters: saving Grandma. The Tiny Lab Association is my last hope to develop a new drug against Alzheimer's disease fast enough.

On the first morning of my apprenticeship, I am nervous but eager. I ground myself with a few deep breaths as I wait for him. He meets me at the door and lets me in. My independent research helps me understand what the tiny labs can achieve, and how I can push them further. However, I still have much to learn.

I have read a lot about tiny labs but could not imagine how small they are. Brad pinches a hair from his head and holds it in front of me. "The tiny droplets are about as thick as this single strand of hair," he explains. Brad then shows me the project of another apprentice running some tests. With the microscope, I can see the small droplets moving around in the channels. They hypnotize me as they jiggle from side to side, mixing their content while trapped.

First, I must build the tiny lab to contain the tiny droplets. The task is not trivial considering their small dimensions. The scientists who created tiny labs used techniques from building microchips. Brad says he'll teach me the two main steps: making the mold, and the tiny device.

Brad thoroughly briefs me on security precautions in the lab. I feel like a mad scientist with my safety glasses surrounded by all the chemicals and equipment. All the manipulations require lab coats and gloves to handle the chemicals. I am Brad's shadow.

He first prepares the mold that will shape the device. I diligently note the equipment and chemicals required for the process. We inspect the completed mold using a microscope. Everything looks all right.

With the mold in hand, we can fabricate one tiny lab. The process is simpler and quicker. However, I am surprised when Brad starts to put things away and clean the lab. I ask him, "Why are we stopping now? The device is unfinished. We need to execute more tests today."

He explains that the device must harden overnight. We cannot rush the process. I attempt to argue for a quick workaround, but none would guarantee the device's integrity. Slightly defeated, I understand that downtime is not wasted time.

After a good night of sleep, I am refreshed and ready to continue building, or fabricating, the device. The device hardened, and we must now separate it from the mold. The base of the mold is fragile. I inadvertently hold my breath. I feel my heart beating. Every setback in the process delays Grandma's medication. Brad slowly and carefully peels away the device from the mold. We will be able to reuse the intact mold. The tiny lab is done.

We inspect it with the microscope to admire our work. Lo and behold, I spot a tiny piece of debris that clogs the device! Disappointed, I ask him to double-check, and indeed, this device is unfortunately compromised. He reassures me that it happens even for more experienced microfabricators like himself. We will reuse the mold to make another device. This will help me get more comfortable with the entire fabricating process.

Day after day, I go to the lab to practice the fabrication process, but I am getting discouraged. We progress steadily but too slowly.

Although I learn so much daily, I get impatient. I need to go beyond this apprenticeship. I ask Brad, and he shows me how he started to code a primitive control system. We need a microscope to look at the droplets. An air pump moves the tiny droplets through the device. A camera watches what happens, and the computer calculates how to use the pump to move the droplets just right.

The control and potential for automation are great. Without any additional sensors (other than the camera), Brad controls the droplet using the computer mouse.

"Look, I can decide to move the droplet here or here," he remarks as he demonstrates to me.

I look at the droplet moving around while thinking, *This is not enough, but it is a good step toward automation for tiny labs.*

"What if the code could do it with one click rather than dragging the droplets manually?" I ask Brad. I continue, "I think I can improve the program with a way to interpret high-level instructions. This would improve the repeatability so much. I have both experience and interest in coding and developing this software!"

He thinks about it, and answers, "Let's get you working on this project right away. I thought we could use your software expertise."

I get excited about moving toward my goal. However, the code development is challenging. I enjoy solving the problems, regardless.

After so many days and devices fabricated, I feel everything slowly falling in place. I select one device from the batch I completed yesterday. The microscope holds the device in place. I hook up the pressurized air pump to push the droplets around. My code loads. The camera-image analysis is done. The calculations are done.

I feel my heartbeat accelerate in anticipation of this fully combined test. Every previous test focused on one single function. I have not tested everything together yet. This is it. I click the right buttons. I enter a few numbers, and I press "Start."

Nothing happens. My heart sinks. However, I notice that I forgot to open the pressure valve. I sigh in relief and chuckle.

The system resets quickly, and now everything should be in place. I press "Start." The computer screen shows the droplet moving around, executing the manipulations that I ordered. I smile. The whole system finally works.

My functional system performs simple medication tests using high-level instructions from which the process is automatically completed. A pharmaceutical specialist must now join the project because I do not know how to develop medication that will work for Grandma.

We bring in Alice—one of Brad's university colleagues—to help us with the medication chemicals.

Alice mentions, "I was aware of tiny labs but could not figure out how to use them."

Alice is amazed at how the tiny lab works. She can't wait to get started. I explain to her how to use it, and she starts working with it right away.

Her experience with conventional labs echoes my own. Each test lasted two hours! No wonder the labs I visited had long wait times. When I tell her the tiny labs complete the test in five minutes, she already starts talking about all the experiments she wants to run and how this will help her.

The first validation test we do, we all stare at the screen waiting for the measurement signal, but nothing happens. After a little while, Alice breaks the silence "Well, that did not work. Let's find out why."

My heart drops. This is not part of the plan. I breathe deeply. I count to five and start to assess the situation.

I question Alice about her procedure for the conventional test. "Light and warmth are the two main enemies here," she emphasizes.

My heart pounds in my chest. We already blocked the light from the solutions during the preparation and put the vials on ice. How can this fail? We cannot get this far and falter. "Would the microscope light we shine also affect the results negatively?" I ask them.

They confirm the possibility, but usually it is not an issue. However, our setup differs from conventional ones. The light could affect the tests more than we initially thought.

Brad and I add a little shutter, like a tiny lid, that opens and closes automatically to block the light during the experiment. We run another test. I look at the screen but do not see the signal. A feeling of discouragement starts to overcome me. Alice squints her eyes, moving closer to the screen.

"Look! A signal, a very faint signal, but a signal!" she exclaims.

My heart jumps, filled with hope. What else could be the problem? Temperature? "The vial is cold, but is the liquid cold in the tiny lab?" I tentatively ask.

Brad thinks out loud, "Usually, the liquid does not warm up. However, this automation handles flow differently."

We quickly calculate. Indeed, the room temperature is enough to warm up rapidly the liquid and limit the signal. My heart beats fast from excitement.

One last problem to solve.

We isolate the tubing with foam and add additional ice for the vial just to be sure. With light and temperature now under control, we hope this test will work.

I reset the system and start the test. Before we even wonder whether it is working, we see the signal. We all celebrate. More tests must be performed, but they will be a breeze with the automated system, making it easy to repeat and validate the medication tests. Alice starts working on a plan for all the tests right away.

CURIOUS GIRL NOTE—Why isn't the tiny lab working?

- Could be too much light affecting the liquid. Is the microscope light too strong?
- Did the liquid get too hot in the lab?
- Controlled the light and temperature. Now it works!

The automated tiny lab system we built has so much potential to help people. My system can help people without experience use tiny labs! With reasonable training, other medication experts like Alice will be able to use my tiny lab autonomously. Their tests will be performed rapidly to explore a lot more of their ideas. Many people from the aging population—including Grandma—will benefit from the new medications that can be available faster and cheaper.

One day, a few years later, I settle next to Grandma after Riley left for the day. "How is your apprenticeship going?" she asks.

"Oh, it ended a while ago, but I stayed in touch with the Tiny Lab Association. The project continues, and I am now a liaison officer," I answer.

These days, many more projects fill my schedule. I know a little about everyone's expertise, so I'm excellent for the job of liaison officer. I am part of a team looking into expanding to other neurodegenerative diseases. I still follow the progress of the project for drugs against Alzheimer's disease from my apprenticeship. Our work led to the development of a few new drugs that passed the preclinical trials. Grandma luckily got selected for the trials. She receives regular doses, and they monitor her response. Thankfully, her condition has stopped getting worse, and she still remembers me! A lot remains to be done, but for now, I get to enjoy a well-deserved *sucre à la crème* with Grandma.

epilogue

So You Want to Be a Fluid Dynamicist?

We have so much more to say about being a fluid dynamicist than these stories could hold. If you're reading this epilogue, we're hoping it's because you're a Curious Girl who is now curious about fluid dynamics. Below, you will find answers to questions that we thought you might have. If you have more, let us know (at storiesofwomeninfluids.org)!

What skills do you need to become a fluid dynamicist?

A good foundation in basic math and physics is essential for learning any kind of engineering and sciences like fluid dynamics. However, you don't have to be an A+ student or a "natural" to develop this skill. Many of us, like Banafsheh and Kelli, weren't the best math students at your age. We worked at it over time to improve.

You might have noticed some common themes in the stories you just read, like teamwork. As researchers, teamwork is part of our daily life. Every one of us is either leading a team or a part of a team—or a part of one team and leading another. The lone wolf researcher is rarely successful.

Being a fluid dynamics researcher also requires patience. We are not always patient people, like when Beverley kept checking if her code worked in "Controlling Eddies." But part of science and engineering is generating new knowledge. This can take a substantial amount of hard work and time.

The most important skill is persistence. You'll note in every story, we are continuously trying, failing, and trying again until we find a way to be successful. This kind of perseverance and resilience is what being a researcher is all about. Every experiment that doesn't give us what we are expecting is a chance to learn something new about our work and ourselves.

Once you make a discovery, what happens next?

Scientific discoveries happen one step at a time. Once we understand one part of the work—like the small jets in Anne-Marie's story "Moving Faster Than Sound!"—it helps us understand other parts of the problem even more. For Anne-Marie, learning how the jets modify the flow to reduce the separation helps her better understand the spacing of the jets, the direction the jets have to be facing, and how strong the jets need to be.

To share this knowledge, we write about our discoveries so that other researchers can read what we have learned and try to answer their own research questions. Sahar found the secret to Hannah's turbine problem in papers written by other researchers in "Mysterious Forces in the Wind."

Multiple times a year, researchers from all over the world come together at scientific meetings and conferences to discuss their work. Sometimes our discoveries get handed off to other engineers who want to take what we have discovered and build it into their designs.

Our discoveries often create practical questions that we have to answer, such as how sensitive are the jets we make to the size of the holes we drill? Or how does the whole air-jet system need to be set up if we put it on a space launcher instead of a wing?

Then there are the questions that exist because they are the next step. Anne-Marie's next step is to understand what happens if she places the jets on a real airplane wing rather than a flat plate. Beverley wants to know how to tickle the turbulence in just the right way to force it to do what she wants. We have so many questions and ideas, and so much to discover in fluid dynamics, that there is always something new to do.

The hardest part about making a discovery is to celebrate that we discovered something before we start working on the next step!

What is it like to do research?

The stories you just read give you a sense of what it is like to do research. There is a lot of "Try, fail, assess, and try again" in research. "That is what research is," as Kelli's PhD advisor was quoted as saying in "Breaking Waves and Derailed Dreams."

One could say, "Research is perseverance," but it is more than that. It is also knowing when to try a different approach or get input from colleagues when you are stuck. And that asking for help is not only okay, it is also necessary for success.

One could also say, "Research is teamwork." We all believe that working with a good group of motivated and curious people is the best part of our jobs.

The best part of research, however, is following our interests—our curiosity—and asking a lot of questions. If all goes well, we even answer our own questions. Doing research in fluid dynamics allows us to follow our passion and contribute to a field that impacts so many aspects of everyday life.

What is it like to be a professor?

Being an engineering professor is an exciting job. Most of us have been, are, or are just becoming professors. The job of a professor often combines many different activities and responsibilities. One well-known responsibility is teaching college courses. We are happiest teaching classes inspired by our research, like fluid dynamics or aerodynamics. But we also teach classes that help us do our research, like engineering design or programming to solve fluid dynamics problems.

However, being an engineering professor involves so much more than teaching. We direct our research labs and work closely with our student researchers. Mentoring our students to help them develop into full-fledged researchers is another favorite part of our job.

Many of us also do outreach efforts, like creating hands-on activities and visiting your classes in the hope of sparking your curiosity about science and math. We want to demonstrate that science is for everyone and that fluid dynamics is everywhere. It is a big reason why we all participated in this book!

What is it like to collaborate with other researchers?

For many students, school group projects can have a pretty bad reputation . . . at least from some of the experiences we remember when we were in school. A lot of group projects went poorly and were full of drama about who was doing the work and whether or not we were going to get it done on time.

We won't lie to you: Collaborating with other researchers can be like a bad group project when the group doesn't come together well. But, did you notice how many people were involved in Christina's "Wonders Flowing in Our Veins" and Marie's "Fluid Mechanics to Grandma's Rescue"?

Fluid dynamics touches so many parts of life that no problem is *just* a fluids problem. Our interdisciplinary projects often require expertise in chemistry, biology, materials, and computer science. Christina's malaria project also needed experts in tropical disease and medicine development! A diverse team coming together with their individual expertise is how we can tackle complex problems.

Successful teams work hard to not be that "bad group project." Teammates have to spend additional time learning how to communicate with each other. This is especially important when team members have different expertise and responsibilities. Teammates take time to agree on project goals and prioritize everyone's ideas.

It takes extra work to be on collaborative teams. But while it can be hard to do and take time to get it right, we believe the rewards of collaborating with a good group of curious people are worth it—which is exactly what this book proves!

What are the different career paths for a fluid dynamics researchers?

We are all fluid dynamics researchers associated with universities and colleges around the world—that is how we met and decided to write this book. Yet there are many career paths in fluid dynamics beyond academia.

With the foundation of math and physics needed to become a fluid dynamicist, any kind of teaching and research are open to you. However, there are many research jobs that need us that you might never have thought of.

For example, have you ever noticed the background hum in your house? It's only when the power goes out that we realize how noisy everything is. Most of those hums are the cooling systems for the electronics in your house. Tech companies need fluid dynamics experts to design cooling systems for all of the gadgets in your house to keep them from overheating.

Beverley and Christina cleverly pointed out that fluid dynamics is even within us—how we breathe and how blood moves through our body. Medical device companies need fluid dynamics experts to design better equipment to help patients breathe and better ways to pump blood through the body.

After reading all of the stories on airplanes and rockets, it should be clear that aerospace companies like Airbus and Boeing require specialists to design airplanes that maximize aerodynamic efficiency, reduce drag, and improve fuel consumption. Fluid dynamicists play a big role in designing launch vehicles for space agencies like NASA and the European Space Agency but also private companies like SpaceX and Sierra Nevada Corporation.

Karen's running journey alongside the development of wind power technology showed how we are now ready to put it to use. As we get more and more wind farms built and operating, fluid dynamicists will be needed to start answering the next practical research questions on wind power like, "Can we build big floating wind turbines in the ocean?" And, "Are there new turbine designs that can capture the energy in a different way?"

Because fluid dynamics is everywhere, the types of jobs we can hold are infinite.

References and Resources

Swimming Faster and Smarter: The Story of a Girl Who Dreams of Going to the Olympics

Hall, Nancy. "What is Drag?" NASA, Glenn Research Center. July 21, 2022. https://www1.grc.nasa.gov/beginners-guide-to-aeronautics/what-is-drag/#:~:text=Drag%20is%20a%20mechanical%20force,without%20being%20in%20physical%20contact.

Reynolds, Gretchen. "Delineating the Perfect Swim Stroke." *The New York Times*. July 4, 2012. https://archive.nytimes.com/well.blogs.nytimes.com/2012/07/04/delineating-the-perfect-swim-stroke/?searchResultPosition=1.

Prince, Andrew. "The Science Behind Swimmers' Dolphin Kick." *NPR*. August 13, 2008. https://www.npr.org/2008/08/13/93566627/the-science-behind-swimmers-dolphin-kick

Mysterious Forces in the Wind: The Story of a Girl Who Solves a Mystery about a Wind Turbine

"CFD computation of helicopter blade dynamic stall." CFD computation performed by Onera solver FAST, post-processed with Cassiopee. Posted August 26, 2016. YouTube. 1 minute, 19 seconds. https://www.youtube.com/watch?v=2ZMKWB3tQV8

"How to Make a Parachute – DIY Easy Plastic Parachute." Video created by Fixitsamo. Posted July 11, 2018. YouTube. 6 minutes, 9 seconds. https://www.youtube.com/watch?v=fnA2X6xLi3o.

"How to make a spinning paper Pinwheel DIY, paper windmill craft." Video created by Simple Art and Crafts. Posted July 14, 2019. YouTube. 2 minutes, 35 seconds. https://www.youtube.com/watch?v=Yth6_O6ShPc

McCroskey, W. J. "The Phenomenon of Dynamic Stall." *NASA Ames Research Center,* Tech. Memo 81264 (March 1981): pp. 1–28. https://ntrs.nasa.gov/citations/19810011501.

Mulleners, K., and M. Raffel, "The Onset of Dynamic Stall Revisited," *Experiments in Fluids* 52, no. 3 (May 2011): 779–793. doi: 10.1007/s00348-011-1118-y.

Le Fouest, S., and K. Mulleners, "The Dynamic Stall Dilemma for Vertical-Axis Wind Turbines," *Renewable Energy*, vol. 198 (October 2022): 505–520. doi: 10.1016/j.renene.2022.07.071.

Facing Strong Headwinds: The Story of a Girl Curious about Wind Energy

Lindenberg, Steve, et al. *20% Wind Energy by 2030: Increasing Wind Energy's Contribution to US Electricity Supply.* US Department of Energy, Energy Efficiency and Renewable Energy, July 2008, chrome-extension://efaidnbmnnnibpcajpcglclefindmkaj/https://www.energy.gov/eere/wind/articles/20-wind-energy-2030-increasing-wind-energys-contribution-us-electricity-supply

Musial, Walter, et al. *Offshore Wind Market Report: 2022 Edition.* US Department of Energy, Office of Energy Efficiency & Renewable Energy. August 2022. chrome-extension://efaidnbmnnnibpcajpcglclefindmkaj/https://www.energy.gov/sites/default/files/2022-09/offshore-wind-market-report-2022-v2.pdf

"Wind." IEA. Last updated February 25, 2025, https://www.iea.org/energy-system/renewables/wind.

"Wind Power Technology Trends." Energy Technologies Area, Berkeley Lab, Lawrence Berkeley National Laboratory, 2025, https://emp.lbl.gov/wind-power-technology-trends.

"Wind Turbines: the Bigger, the Better." US Department of Energy, Office of Energy Efficiency & Renewable Energy. August 21, 2004, last updated August 2024. https://www.energy.gov/eere/articles/wind-turbines-bigger-better

Wiser, Ryan, et al., *Land-Based Wind Market Report: 2021 Edition.* US Department of Energy, Office of Energy Efficiency & Renewable Energy. August, 2021, chrome-extension://efaidnbmnnnibpcajpcglclefindmkaj/https://www.energy.gov/sites/default/files/2021-08/Land-Based%20Wind%20Market%20Report%202021%20Edition_Full%20Report_FINAL.pdf

Acknowledgments

This anthology is the result of a shared vision and the collective efforts of many talented volunteers. I am deeply grateful to everyone who joined me on this journey of bringing this book to life.

First and foremost, my deepest thanks go to the members of the SOWIF Leadership Team who have contributed their wisdom and effort to this project: Dr. Roni Goldshmid who joined as a postdoc at California Institute of Technology and is now an assistant professor at San Diego State University; Dr. Swathi Krishna, assistant professor at University of Southampton; Dr. Beverley McKeon, professor at Stanford University; Dr. Banafsheh Seyed-Aghazadeh, assistant professor at the University of Massachusetts Dartmouth; and Dr. Nicole Xu, assistant professor at the University of Colorado Boulder.

To the contributing authors to this anthology, thank you for sharing your unique voices and perspectives. Your trust, time, and patience—especially as we learned the process alongside you—are greatly appreciated. Your experiences form the heart of this collection, and it has been a privilege to bring them to light.

Dr. Karen Flack (US Naval Academy, Retired), Dr. Elizabeth Gregorio (George Washington University), and Dr. Christina Rice (University of Michigan) joined members of the SOWIF Leadership Team (myself and Beverley) to form the editorial board for this anthology. To this editorial team, thank you for ensuring that the stories in this anthology were told with wonder and heart.

To Cosmic Writers, particularly Rowana Miller and Nancy Xu, thank you for helping our authors pivot from experienced technical writers to novice storytellers. The expertise you shared and the care you brought to this process were invaluable. You helped us grow as writers and communicators, and we are deeply grateful for this relationship.

Jane Shore brought her amazing talents to the illustrations within the chapters in this anthology. Thank you for sharing your gifts and bringing us to life.

Roni Goldshmid provided the scientific context as Jane translated our scribbles into the beautiful science illustrations.

Beverley McKeon and Nicole Xu drew on their inner Curious Girl to craft the text for the *Curious Girl* Notes contained within these pages.

To my virtual assistant, Sheri Hooton, for keeping my spirits up and helping with so many of the things that kept falling off my plate.

To She Writes Press, who brought our books into the Fall 2025 cohort. Thank you for your infinite patience and guidance as we navigated the challenges of finalizing these manuscripts during the end of an academic semester.

This spark of inspiration for this book would not have happened without Dr. Tiffani Teachey and the initial starting materials on how to write a professional narrative piece that she supplied us.

In the two-year journey of taking this project from ideation to publication, SOWIF has benefited from many funding resources, support, and Friends of SOWIF contributions.

Funding resources:

- The 2023 American Physical Society (APS) Forum on Outreach and Engaging the Public Mini-Grant provided seed funding, which initiated our partnership with Cosmic Writers.
- Physical Review Fluids, which provided travel support for Nicole Sharp and Rowana Miller to participate in the 2023 APS Division of Fluid Dynamics SOWIF Minisymposium.
- The Caltech Postdoctoral Association provided early funds to support initial developmental editing with Cosmic Writers for participating Caltech postdocs.
- The University of Colorado Boulder's Mechanical Engineering Department's Diversity, Equity, and Inclusion Working Group provided funds supporting Dr. Nicole Xu in raising the visibility of this project.
- Office of Naval Research (ONR) STEM grant (Award N00014-24-1-2043), under the supervision and guidance of Dr. Yin Lu (Julie) Young, moved this project to publication with a critical funding injection. It enabled the majority of the editing, illustration, and publication of this book with She Writes Press.

Support:

- Dr. Anette "Peko" Hosoi (APS/MIT) for support and guidance in creating the APS DFD Minisymposium.
- Dr. Thomas Fu (ONR) guided us to an appropriate funding call that could support the publication of this anthology.
- Our ONR Program Officer for the STEM grant, Dr. Yin Lu (Julie) Young, continuously worked to raise the visibility of this project as we brought it to completion.

Friends of SOWIF:

- Professor Bharath Ganapathisubramani at the University of Southampton, whose early support enable us to continue the partnership with Cosmic Writers.
- PeakCells's design of the logo and brand colors that captured the spirit of SOWIF and gave us an identity we didn't know we needed until we saw it.

Out of a desire for full transparency, I want to acknowledge my use of ChatGPT, specifically the Professional Writing Coach model authored by ChatGPT, to structure and fine-tune the preface, introduction, and acknowledgments of this anthology. The Creative Writing Coach model authored by ChatGPT provided additional editorial support, especially in identifying and crafting age-accessible language in our science communication. When it comes to new technologies, it can often be unclear whether to embrace them or keep them at arm's length; for me, though, ChatGPT has served as a mentor and tutor as I continue my journey from technical writer to whatever it is I am now.

About the Authors

Anne-Marie Schreyer

Anne-Marie is currently the leader of a research group at RWTH Aachen University, Germany. Her group uses experimental and numerical methods to study transonic and supersonic flows with unsteadiness, including transonic airfoil buffet and shock-wave/boundary-layer interactions. Further topics of Anne-Marie's research are the control of separated flow and fluid/structure interactions. Anne-Marie is an aerospace engineer and has a doctoral degree in fluid dynamics, both from the University of Stuttgart in Germany. As a postdoc, Anne-Marie has worked at Aix-Marseille University in Marseille, France, and at TU Braunschweig, Germany. Anne-Marie enjoys learning languages, getting to know other cultures, and working together with interesting people. She has therefore spent time abroad to study (Linköping University and the Royal Institute of Technology KTH, Stockholm, Sweden), carry out research (Aix-Marseille University, France; Princeton University and Stanford University, USA), and travel (Portugal, France, United Kingdom, Sweden, Norway, Denmark, New Zealand, Nepal, etc.).

Banafsheh Seyed-Aghazadeh

SOWIF Leadership Team

Dr. Banafsheh Seyed-Aghazadeh is an assistant professor of mechanical engineering and directs the Laboratory for Fluid-Structure Interactions Studies at the University of Massachusetts, Dartmouth. She has received awards, including the prestigious 2022 National Science Foundation Faculty Early Career Development Program (NSF CAREER) award, the 2022 Office of Naval Research Young Investigator Program (ONR YIP), and UMassD's College of Engineering Faculty Award for Diversity, Equity, and Inclusion. Previously, she held the James R. Myers Endowed assistant professorship at Miami University and worked as a postdoctoral research associate and lecturer at UMass Amherst where she also earned her PhD in mechanical engineering and was recognized as the Outstanding Young Alumna in 2021. Her research focuses on experimental fluid-structure interactions (FSI) with applications in biomechanics, aerospace engineering, and energy systems. She also promotes outreach activities for underrepresented minorities in early education, encouraging young girls to pursue careers in STEM.

Beverley McKeon

SOWIF Leadership Team, Anthology Editorial Board

Beverley McKeon's research interests focus around the understanding and control of turbulence, a physical phenomenon that exists in flows on many scales throughout daily life: in rivers, atmospheric, and biological flows; around cars, aircraft, and ships; and inside large-scale (water and natural gas) pipelines. Learning how to describe and model turbulence is her passion and also (very fortunately) her job. Her work has been recognized as advancing the state of the art in her field and has led to a range of leadership roles in the United States and international scientific communities. Beverley is a professor of mechanical engineering and member of the Center for

Turbulence Research (CTR) at Stanford University. She moved to Stanford in 2023 after sixteen years at Caltech, where she was the Theodore von Karman Professor of Aeronautics in the Graduate Aerospace Laboratories at Caltech (GALCIT). She received undergraduate degrees from the University of Cambridge and a PhD in mechanical and aerospace engineering from Princeton University before settling with her family in California.

Christina Rice

Anthology Editorial Board

Christina Rice is a PhD candidate at the University of Michigan. She was born in North Carolina and is currently living in Thailand, where she is completing the final portion of her dissertation in collaboration with scientists at Chiang Mai University while also studying the Thai language. She has completed a BSE in mechanical engineering and an MSE in chemical engineering. She has taught engineering manufacturing courses and facilitated at the Center for Socially Engaged Design. Beyond engineering, she has worked with the Gerald R. Ford School of Public Policy to pursue public policy translation. Her community building and science outreach have been recognized through the Martin Luther King Jr. Spirit Award and the Charles Vest Award. Her academic merit has been acknowledged through prizes, including the Enoch J. Durbin Prize for Engineering Innovation, and grants, including the National Science Foundation Graduate Research Fellowship. When relaxing, Christina enjoys hiking, listening to music, and spending time with family.

Elizabeth Gregorio

Anthology Editorial Board

Dr. Elizabeth Gregorio swam competitively for eleven years, from sixth grade until she graduated from college. Her proudest moment in swimming was standing on the podium with her friends for three relays at championships her senior year of college. While in college, she also majored in physics (and minored in math) and conducted several independent research projects on ways that she and her teammates could swim faster. After college she went on to pursue a PhD in mechanical engineering, studying why Olympic divers don't make a splash when they enter the water. Because of swimming, she discovered a passion for understanding the fluid dynamics of biolocomotion, which she is continuing to study as a research scientist in Paris. She was able to reach her dream of going to the Olympics, watching in person in Paris 2024. Becoming a swimming scientist ended up being the perfect "backup plan."

Karen Ashby Flack

Anthology Editorial Board

Karen Ashby Flack is a professor emeritus in the Mechanical Engineering Department at the United States Naval Academy in Annapolis, Maryland. She received a bachelor's degree from Rice University; a master's degree from the University of California, Berkeley; and a PhD from Stanford University—all in mechanical engineering. During her tenure at the US Naval Academy, she taught thermodynamics, fluid mechanics, heat transfer, design, and wind energy courses. Her service to the institution, which included a term as department chair, was recognized with a US Government Superior Service medal. She is a Fellow of the American Physical Society citing her experimental research on the physics of rough-wall turbulent boundary layers. Currently, she is a research faculty member in the Energy Academic Group at the Naval Postgraduate School, coauthoring an interdisciplinary book on energy security.

Kelli Hendrickson

SOWIF Founder, Anthology Editorial Board

Kelli Hendrickson, ScD, PCC, earned her Doctor of Science in ocean engineering from MIT in 2004, focusing on air-water interactions during wave breaking through computational simulations. She served as a research engineer in MIT's Department of Mechanical Engineering, leading research as coprincipal investigator on multiple grants to model multiphase air-water turbulent flows, until 2024. Now a research associate professor at UNC-Chapel Hill in the Department of Mathematics, her research centers on multiscale modeling of turbulent air-water flows and biofluids. In 2019, Kelli began her path in career and leadership coaching, becoming certified in 2020 and attaining her Professional Certified Coach (PCC) credential in 2024. She is also the founder and principal instigator of the Stories of Women in Fluids (SOWIF) Initiative, committed to amplifying the voices of women in STEM. Her unique background in research and coaching enables her to empower STEM professionals to navigate and thrive in their careers.

Leah Mendelson

Leah Mendelson is an associate professor of engineering at Harvey Mudd College. She directs the Flow Imaging Lab at (Harvey) Mudd (FILM), leading research projects focused on the hydrodynamics of aquatic locomotion, bio-inspired mechanism design, and imaging techniques for fluid flow measurement. Professor Mendelson also regularly teaches courses in mechanical design, solid and fluid mechanics, and introductory engineering. As an educator, she emphasizes experiential, student-centered, and interdisciplinary approaches to mechanical engineering. She holds a BS from Olin College (2011) and an MS (2013) and PhD (2017) from MIT, all in mechanical engineering. Outside of the lab, her hobbies include hiking with her dog, jigsaw puzzles, and (unsurprisingly) swimming.

Marie Hébert

Marie Hébert earned their PhD degree in mechanical and mechatronics engineering from the University of Waterloo

(Canada) in 2020, in addition to a graduate studies Certificate in University Teaching. She received several scholarships such as the Nanofellowship from the Waterloo Institute for Nanotechnology and the Canada Graduate Scholarships-Doctoral (CGS-D) from the Natural Sciences and Engineering Research Council of Canada (NSERC). Originally from the province of Québec, Marie speaks both French and English. She previously completed a bachelor's of mechanical engineering degree with a mechatronics specialization at Concordia University (Canada). She received the Mechanical Engineering medal for her top academic performance. As an assistant professor at the Université du Québec à Trois-Rivières (UQTR), their research focuses on both fluid dynamics and mechatronics—active microfluidics. Their research interests include multiphysic system modelling, control systems, microfluidics, fluid dynamics, and fuel cell systems. Marie is also an outdoor enthusiast.

Nicole Xu

SOWIF Leadership Team, Illustration Editor

Dr. Nicole Xu is an assistant professor in the Paul M. Rady Department of Mechanical Engineering, Robotics Program, Biomedical Engineering Program, and BioFrontiers Institute at the University of Colorado Boulder. Prior to joining CU Boulder, she was a National Research Council Research Associate in the Laboratories for Computational Physics and Fluid Dynamics at the US Naval Research Laboratory. She received her PhD in bioengineering from Stanford University, her MS in bioengineering from the California Institute of Technology, and her BSE summa cum laude in bioengineering from the University of Pennsylvania. Her research group focuses on developing bioinspired robots and underwater vehicles for applications in environmental monitoring, such as tracking markers of climate change. Her work has been featured in national and international media outlets, including BBC World

News and various TV, radio, and podcast appearances as well as in print publications such as *Wired*, *Scientific American*, and *The Economist*.

Ritambhara Raj Dubey

Ritambhara Raj Dubey was born and raised in Raipur, a small city in central India. After completing her bachelor's and master's degrees in mechanical engineering, she gained experience in the automobile industry before pursuing her PhD at Florida State University (FSU). Her research centers on fire science, specifically focusing on the interaction between fire and the atmosphere. Outside her academic endeavors, Ritambhara enjoys swimming, hiking, and dancing. She is passionate about continuing her journey in academia as a researcher, aiming to blend her love for teaching, research, and community service to make a meaningful impact within academia and the broader community.

Roni Goldshmid

SOWIF Leadership Team, Illustration Editor

Dr. Roni Goldshmid is an assistant professor of aerospace engineering at San Diego State University. She holds a BS from the University of California, Berkeley, and an MS and a PhD from the Technion-Israel Institute of Technology. A former Caltech Presidential Postdoctoral Fellow and a Rising Star at Stanford University, Dr. Goldshmid is the recipient of the Grinshpen Prize for Excellence in Environmental Engineering. Her research focuses on fluid structure interactions, heat transfer, and flow control to address the complex challenges of aerial wildfire suppression. Her current interests include developing advanced experimental and theoretical models, creating predictive statistical tools for wind-fire-vegetation interactions, and designing data-driven optimization algorithms for aerial firefighting operations. By integrating fundamental scientific principles, advanced machine learning techniques, and cutting-edge aerodynamic systems, Dr. Goldshmid aims to enhance the safety, efficiency, and effectiveness of aerial firefighting operations.

Sahar Rezapour

Sahar Rezapour is a PhD candidate in mechanical engineering at EPFL in Switzerland. During her bachelor's and master's studies, she developed a passion for fluid mechanics in energy conversion technologies. Her PhD research investigates the dynamic stall phenomenon—a significant aerodynamic challenge encountered in various fields, including micro-air vehicles, helicopter rotors, and wind turbines. She employs experimental methods to analyze the repeated patterns in this phenomenon, aiming to understand the underlying physics and improve their predictability. In 2023, Sahar was awarded the Amelia Earhart Fellowship offered by Zonta International for her academic achievements in aerospace sciences. Eager to promote science, she has joined the organizing committee of a seminar series at EPFL to facilitate knowledge sharing. She was also among the EPFL finalists of the My Thesis in 180 Seconds (MT180) competition. Through her work, she hopes to inspire future generations to explore the world of STEM.

Shilpa Vijay

Dr. Shilpa Vijay is a postdoctoral researcher at Stanford University's Center for Turbulence Research. She completed her PhD in mechanical engineering at the University of Southern California, with a focus on passive flow control using porous surfaces. She holds a BTech in civil engineering from the College of Engineering, Pune, in India. Shilpa is a curious individual whose diverse interests—ranging from sports and cooking to comedy—provide fresh perspectives on her scientific work. She enjoys finding new insights in everyday activities and believes strongly in the unique journey each individual takes in science. Her passion for learning extends to her hobby of quizzing, which deepens her curiosity and encourages creative thinking. Through both her research and mentoring, Shilpa is committed to supporting the growth and development of the next generation of researchers while fostering an appreciation for the diverse paths that shape each person's scientific journey.

Swathi Krishna

SOWIF Leadership Team

Dr. Swathi Krishna is an assistant professor of experimental aerodynamics and propulsion at the University of Southampton, United Kingdom. She earned her bachelor's degree in mechanical engineering from Visvesvaraya Technological University, India, in 2009; followed by a master's in aerospace engineering from TU Delft, Netherlands, in 2012; and a PhD in mechanics from École Polytechnique Fédérale de Lausanne (EPFL), Switzerland, in 2017. After a brief stint in industry, she conducted postdoctoral research at the University of Rostock, Germany, before joining Southampton's faculty in 2021. Dr. Krishna's research focuses on unsteady, vortex-dominated flows, drawing inspiration from nature to enhance performance metrics in engineering systems. Her team develops robotic systems for fluid environments, combining experimental techniques and flow diagnostics to understand fluid-structure interactions. These insights inform design and optimization, with applications in flapping wings, vertical-axis rotors, and propeller-wing interactions. In addition to her research, Dr. Krishna is dedicated to outreach, regularly leading programs for local students aged ten to eleven and fifteen to sixteen to inspire the next generation, especially young girls, to pursue STEM careers.